The
Trainspotter's
Notebook

www.penguin.co.uk

Francis Bourgeois

The Trainspotter's Notebook

bantam

TRANSWORLD PUBLISHERS

Penguin Random House, One Embassy Gardens,
8 Viaduct Gardens, London SW11 7BW

www.penguin.co.uk

Transworld is part of the Penguin Random House
group of companies whose addresses can be found at
global.penguinrandomhouse.com

First published in Great Britain in 2022 by Bantam
an imprint of Transworld Publishers

A CIP catalogue record for this book is available from the
British Library.

ISBN 9780857504722

Designed and typeset in Century Schoolbook Pro 11.5/16pt,
Neue Haas Grotesk Text Pro 10/16pt by James Alexander
at Jade Design

Printed and bound in Great Britain by Clays Ltd,
Elcograf S.p.A.

The authorized representative in the EEA is Penguin Random
House Ireland, Morrison Chambers, 32 Nassau Street, Dublin
D02 YH68.

Penguin Random House is committed to a sustainable future
for our business, our readers and our planet. This book is
made from Forest Stewardship Council® certified paper.

To Amy, my girlfriend. Thank you for being there with me on this journey.

Calling at

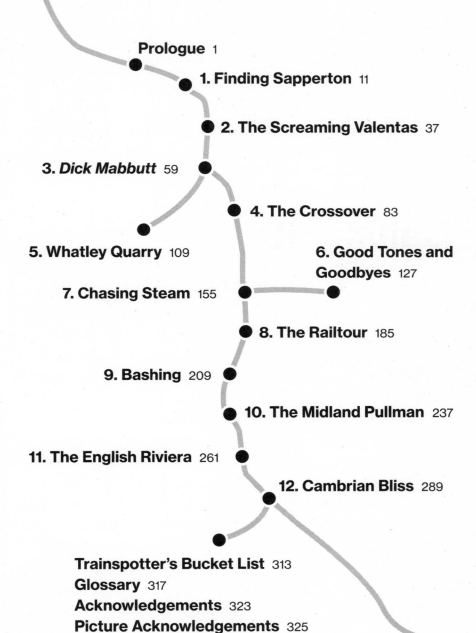

Way in →

Prologue

Visit any railway station in the UK, especially on the main-line, and have a look at the end of the platform. Almost certainly there will be a trainspotter. Sometimes there will be a crowd, at other times just a handful, depending on what's coming through. Trainspotters and train enthusiasts can be any age, any gender, any temperament. It's easy to identify us: a notepad, a camera, sitting on a bench waiting for a locomotive to pass through, standing at the end of the platform setting up a shot. 'Are you here for the railtour?' – a question that can turn a nomadic outing into an adventure with a crew of fellow enthusiasts. Trainspotters like me can find instant camaraderie through our mutual appreciation for the iron horses that gallop across our network.

In the UK, trainspotting began in the early nineteenth century when rail travel was just emerging. The act of being a trainspotter wasn't necessarily recorded back then, but no doubt there would have been wide-eyed children, quietly appreciative parents and curious grandparents marvelling at George Stephenson's *Rocket* as it chugged along at 30mph. By the twentieth century, the romance of travelling by train and the beauty of steam locomotives began to fuel a fascination with the railway. The *Duchess of Hamilton*; the *Flying Scotsman*; the class A4s, particularly the *Mallard* that, in 1938, broke the record for the fastest steam locomotive, achieving 126mph just south of Grantham in Lincolnshire, all brought the bold image of the railway into the minds of the general public and ignited the beginning of trainspotting as we know it.

Subsequently, the numbers on the sides of locomotives, not just the hero engines but the workhorses of the shunting yards, introduced a new dimension to trainspotting in this era. Nearly twenty thousand steam locomotives ran along the network towards the end of the 1930s, all with unique numbers, some with names, to be collected by trainspotters, along with other relevant information, and compiled in a book. Just the same way someone might collect stamps or coins.

After the Second World War, the railway industry went through drastic changes: diesel locomotives started to take over from their steam counterparts, compounded by the modernization plan of 1955, whereby manufacturers were commissioned to create multiple classes of electric and diesel locomotives in small batches. Through trialling on the network, the most reliable, powerful and efficient locomotives were selected for further orders. This created a hive of separate classes of diesel locomotives operating in different regions with varying degrees of rarity, manufactured by individual companies with different engines, while steam was shouting its last hurrah. It was certainly an exciting time to spot trains. Lines of children would sit with their legs over the edge of the platform, all with notepads, writing down numbers, while older spotters stood behind. Through the 1970s and 1980s, electric and diesel locomotives and multiple units dominated the lines. Steam was no longer the principal source of power. Railway photography became an increasingly popular way of cataloguing locomotives that had been spotted.

The 1990s, the era of privatization, brought new liveries. Spotters were now capturing moments on camcorders, though numbers were still being taken down in trusty notebooks. In the early 2000s, the diesel and electric relics of the previous century started to fall away, replaced by more efficient, less polluting and quieter locomotives and multiple units. Some of the original trainspotters started to fall away too.

Trainspotting found a home on social media sites like Flickr and Facebook in the 2010s: a game-changing way to share the rare movements that might have taken place on a particular day. Instagram enhanced this, then TikTok. A new, younger generation of trainspotters have started to emerge and connect with one another. The station platforms of the last century are now virtual, with enthusiast groups, spotting group chats and social media feeds bringing enthusiasts together, sparking new interests and sharing the beauty of the railway with an expanded audience. The platforms are full, and I am glad to be part of it.

Different parts of my brain are tickled by the railway. The huge amounts of momentum in a moving train, juxtaposed by the directional restriction of the rails, gives me an odd sense of calm. There is so much predictability in the direction of the train; it's clear what is going to happen next. Trainspotting also taps into my desire to collect things, which began with Hot Wheels cars when I was a toddler and, later, noting down details of cool cars that I had seen.

Nowadays I record my locomotive encounters in videos and photographs, which I then order and date. The main stimuli from the railway for me are the sounds: screeching rails, vibrating metal, groaning traction motors, thrashing diesel power units, explosive tones. Sonic energy from the railway gives me goosebumps and electricity in my body. I can happily ride along and listen to a class 455 with its refurbished Vossloh Kiepe traction motors all day, swinging up and down with the varying pitch of the motors. The tunnel just after Leatherhead is a particularly great spot for an exhibition of the animal-like scream from beneath the carriages. What may just be the ambience of a commute, for me is an anthem for my passion.

I have written this book to present trainspotting in its totality. Complete moments of elation. Dampening moments of disappointment. From chasing trains into the depths of Wales to reconciling the status quo of cool with my lifelong hobby. I want to give an insight into trainspotting, which from the outside might seem weird, even potentially puzzling as to why someone would stand at the end of a platform or on an overbridge for hours, or drive hundreds of miles in the lashing rain or beating sun, just to see a train.

I hope this book will also help remind people of their childhood passions and, maybe, if they are lying dormant, to reconnect with the joyful feelings of youth. And for fellow trainspotters reading this, particularly those at school, hold on to your passion. The general perception of what is cool may weather your connection to the hobby, but keep it true to yourself. Trains are cool!

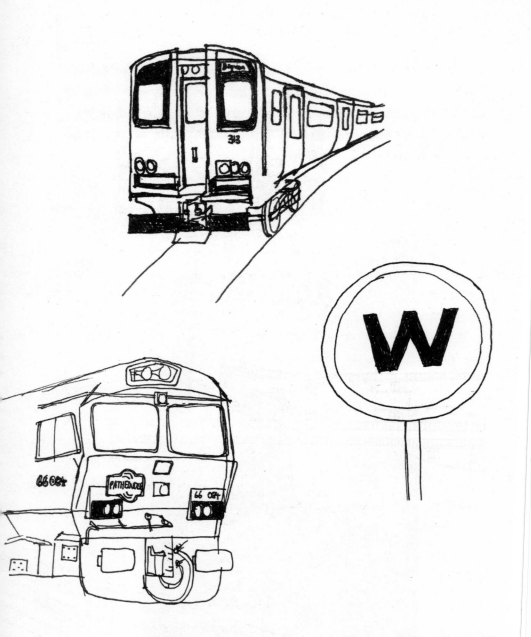

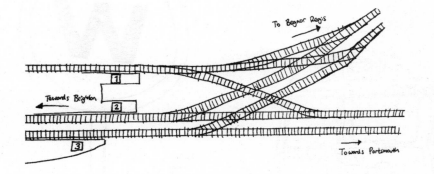

To Bognor Regis

Towards Brighton

1

2

3

Towards Portsmouth

Chapter One →

Finding
Sapperton

Location: A417, Gloucester, 29 December 2021, 21:14

I have stayed in this hotel once before. It's right next to the A417, and just next to the A417 is Barnwood Junction, where the train line coming from Gloucester Station and, further afield, Cardiff meets the line from Bristol Temple Meads. The two lines converge and continue north towards Birmingham.

This is the reason I've chosen this hotel. I can drift off to sleep to the sound of a class 66 hauling the Margam TC to Immingham Sorting Sidings service. As the locomotive joins, the driver will accelerate to notch 8 and the full 3,000hp of the EMD two-stroke engine will bellow from its lungs. Loud enough for me, just audible through the walls.

The last time I stayed in the hotel, I was given a room on the first floor overlooking the car park, away from the line. This time I requested to have a view of the train tracks. In the morning I know I will see two class 37s top-and-tailing a set of barrier coaches on the way to Lydney, to pick up a class 360 that had been used for filming for a TV programme, *Holby City* apparently, three weeks earlier. The class 360 unit was originally meant to have been picked up two weeks ago. I went to see this move, but after watching the start-up and coupling at Lydney Junction platform, there seemed to be brake pressure issues, so the class 37 (37510) returned to Derby RTC light engine. In other words, it was travelling as a locomotive by itself.

The hotel lift takes me to the third floor. The room is just like the one that faced the car park: PVC and laminate flooring. The window view and the mattress are what really matter. I am just falling asleep as the 23:33 Cardiff Tidal TC to Masborough FD blasts by. I get up and pull the railed curtain aside to see the ribbed containers flicker through the illumination of the nearby streetlights. I get back under the rigid sheets after it passes.

I wake up at 05:50. It is still dark.

I check live.rail-record to see where the two class 37s are. They are approximately one hour away. This move is a particularly interesting one as the two class 37s being used are from two separate rail freight operating companies. I get dressed and dash out of my room, remembering my wallet,

Notches

What is a notch? Unlike a car, most locomotives and multiple units don't have a smooth accelerator, they have notches. With class 66s for example, there are eight notches that allow the driver to modulate the power according to how much tractive force is required and the conditions of the railhead. 1 is the lowest power, 8 is the highest. Usually, upon acceleration, shoving the engine into notch 8 will cause wheel slip, so the power must be eased through the notches gently.

my keys, my headphones, my phone and my GoPro.

I arrive at Gloucester Station. My friend Ryan has messaged me on Instagram to say that he is already at the
platform. Ryan and I constantly message about trains, send
each other pictures and discuss our outings. I became friends
with him after he discovered my train videos online. He too
is an enthusiast who likes to film trains. We connected and
arranged our first outing, and since then our friendship has
grown. I have become used to Ryan wearing his bright blue
jacket, so I am looking out for it when I arrive.

The sliding doors of Gloucester Station open. I am slightly
concerned about who is at the ticket gate. Sometimes they
can be a little bit grumpy and not let people through who
want to trainspot on the platform without using a train
ticket.

A very smiley lady greets me and lets me through. Straight
away I see Ryan. He is watching two class 170s uncouple at
the station. It's quite an interesting mechanical process to
witness. Two metallic hands with interlinking fingers able to

hold one another through thousands of newtons of force, just releasing with a clunk.

'Hello, Ryan, how's it going?' I say as I approach him.

'Alright, Francis,' he replies. Ryan is from Gloucestershire and has a very strong accent. He very proudly calls it his farmer accent.

Ryan gets a little bit nervous when meeting new people. I can tell he still gets a bit nervous when meeting me. He's a little bit more rigid and quiet when we first say hello. After talking about the day ahead and what trains we are going to see, he loosens, and his infectious energy starts to roll out. This has a huge impact on me. When I see him, my morning grogginess disappears and it's almost as though my body is anticipating the serotonin that is about to be released into my bloodstream. Every time we go trainspotting together, we have such a great time.

We walk down Gloucester's 602.69-metre-long platform, the longest in England, to the end of the concrete walkway that juts out into the darkness, lit by dim LED lights every 20 metres. We both check the train-tracking website to see that the double class 37s are only ten minutes away. We set up our cameras ready to catch the moment they pull through the station.

At 07:14, 37510 leads two barrier coaches and the 37116 over Horton Road crossing. We can hear the mighty English Electric traction motor even before it's in view of our cameras. With each rotation of the sixty-year-old engine, thunder farts out from its exhaust pipe. A class 66 or any more recent locomotive doesn't come close to the theatrics of this beautiful machine.

The sound of the engine courses through my nervous system, tingling my fingers and sending tickles up my back.

It's almost as though it's a primeval reaction to a dinosaur roaring, fight or flight, but in this case it's simply a tremendous stimulus that makes me giddy.

37510 has recently had a new paint job and it looks absolutely stunning with its yellow, angular snowploughs cutting through the dim, dark blue of the morning. It coasts towards us.

I am expecting nothing more than to see it roll by. Maybe a little tone. It's very unlikely to thrash – which is when the driver opens up the taps and the engine achieves maximum revolutions per minute (rpm), producing the biggest noise it can – or even let off more than two tones, because it's so early. Most of Gloucester is still asleep.

A flatulent torrent from the low tone howls down the platform and it makes me jump. British train horns have two tones, high and low, and in some cases a loud or soft

setting. The driver holds on to the low tone even longer, for approximately two seconds. My jaw drops as I am completely surprised by what I'm hearing. The driver then raises the horn lever and blasts the high tone like Louis Armstrong pursing his lips together and blowing into his trumpet as hard as he can. A sphincter-clenching ring resonates in my ears as it approaches. The driver releases the tone lever. Silence. It's 10 metres away and approaching at 20mph. Five metres. The low tone punches me again, this time louder, closer, more fantastic. I am now laughing uncontrollably.

The driver holds the tone as the locomotive passes me, creating an awesome Doppler effect, where the direction of travel of the sound emitter causes the frequency of the sound to bunch up as it approaches the receiver – me – causing a higher pitch; as it passes, the frequency is extended by the emitter moving away from the receiver, causing a lower pitch. This is one of the ultimate sounds to catch while train-spotting and it's utterly euphoric.

Ryan and I are left on the platform bent over double with hysteria. We high-five as we recover and watch back the footage.

'This is gonna make a great video!' exclaims Ryan. I laugh and nod in agreement.

Tones/horns

Horns can be sounded from 06:00 a.m. until midnight and used to alert trackside workers, when approaching foot crossings or whistle boards, operating in depots or sidings and during wrong-direction movements.

We mill around on the platform for another fifteen minutes and eventually start walking back to the station building.

'I wonder where we can catch it again,' I say to Ryan. 'It's going down to Lydney to collect that 360 so it's going to have six hundred tons or so to haul on the way back to Bicester. We should try and catch it where there's a decent incline so we can hear it thrashing.'

'How about Over crossing?' Ryan suggests.

'Yes, that'd work as you've got that nice incline under the footbridge before you reach the crossing.'

'No, no, no, not that one. I'm not going there,' Ryan cuts in. 'I'm not going back there. That farmer is awful. He threatened to call the police last time.'

Ryan went to the furthest crossing at Over – the third one along – a few weeks ago. It's on a public footpath but for some reason one of the farmers really doesn't like trainspotters.

'I wouldn't be surprised if he shot at us,' Ryan continues. 'Last time I heard him shooting birds, the nasty man.'

Ryan suggests that we go to the first crossing at Over but I know that section of the track is on a decline, so the engine definitely won't be thrashing.

One of the things I find most exciting about trainspotting is discovering new viewing locations. I open Google Earth on my phone and trace the track the two class 37s dragging the 360 will be following, while also looking at the elevation change and any foot crossings. It's quite a hard task that often requires full concentration and some inside intel too.

As we wander towards the lift, we see a driver in Great Western Railway uniform. He looks to be in his mid-thirties. He smiles. I smile back and then look down at my feet.

'Francis?' The driver is still looking at me with a smile on his face.

I reply, 'Yes, how's it going?'

'I love your videos, man. My name's Colin. I'm a driver for GWR.'

'Nice to meet you, Colin. Thanks!' I say with a smile. 'What trains do you drive?'

'Ah, mainly class 800s from Paddington, Cardiff Central and Bristol Temple Meads.'

We talk for a little while about his experience on the railway. He is actually the youngest person ever to drive my favourite train – a class 43 HST. When he left school, he went straight into an apprenticeship and was in the driver's cab by the time he was twenty. Needless to say, I'm jealous.

Ryan butts in. 'We just saw two class 37s heading to Lydney.'

Colin turns to Ryan. 'Oh! That was you, was it? Heard you from a mile off.'

We all chuckle.

I look around as the conversation dries a little, so I decide to tell Colin that we are looking for a good incline to see the 37s on their return journey.

'If it's going back to Bicester via Stroud you should head to the Stroud Valley. Loads of great spots down there and it's an incline from Stroud all the way to Sapperton Tunnel.'

Colin seems adamant that the Stroud Valley is the right spot, and it immediately piques my interest. I have a look on Google Earth and see that there is a succession of lighter areas of ballast every so often along the line, highlighting where the foot crossings are.

'There was a death there last year, so they might've shut some of the crossings,' Colin tells us. 'Network Rail wants to do away with all foot crossings and road crossings as they pose such a safety concern.'

Stop
Look
Liste
Beware
of train

There's a chance all crossings will be shut, but it's a chance I am willing to take. I can see Ryan is worried. He likes to know what's going to happen, as do I to an extent, but I feel for Ryan because it has more of an impact on his mood and emotions.

I check the time. 'What would be the nearest station to the Stroud Valley incline?' I ask Colin.

'Well, you've got Kemble one side and Stroud the other side,' Colin replies.

I check live.rail-record to see that it's due through at 12:50, giving us nearly five hours. 'Ryan, we have bags of time! We might as well check it out.'

Reluctantly, Ryan agrees.

Location: Frampton Mansell, 30 December 2021, 08:52

We jump in the car. I have the destination on my Google Maps set to Frampton Mansell, which seems to be a sleepy village in the Stroud Valley. Only a church, the Crown Inn and the village hall are highlighted on the map as places of interest. The train line runs right through the village.

Google Maps announces that we have arrived at our destination. I am careful not to scratch my alloy rims as I bring the car towards the roadside. Most modern kerbs are tapered, so the tyre will scrub against them before the rims do, but I can see this is not the case with this decrepit pavement. The kerbstones stick their abrasive chins out at my wheels. I leave a good 20 centimetres between them – probably too much.

Ryan and I get out and are immediately knocked sideways

by the sourness of the air. It's not quite the Cheddar-like smell of cow dung. More citric.

'What the hell is that, Francis?' Ryan's face is screwed up, his nostrils flared.

I laugh. 'Smells like shit.'

I decide to swap my grey, crinkled Crockett & Jones loafers – a charity shop find – for some sturdy steel toecaps that I have in my boot. Their glossy black faux-leather finish clashes with my mid-wash Levi's and my blue knitted jumper. I lace them up, shut the boot and lock the car.

Ryan has his big green camping backpack with him carrying an assortment of tripods, portable chargers and energy drinks.

We head towards the green footpath sign next to St Luke's Church, passing a 1960s bus stop. It's constructed from limestone but is not made very prettily, the lichen on it looking quite convincingly like pressed chewing gum.

As we get nearer to the church, the road dips down, revealing a small septic tank truck doing its business with a small cottage.

'That's the smell, Ryan,' I chuckle.

The footpath takes an abrupt right from the pavement and leads down, next to the graveyard. Overhanging branches and brambles tease us as the tarmacked path gets crumblier and steeper. 'This path should take us to the first foot crossing,' I explain to Ryan. The path turns to rubble and I start to see the distinct shape and colour of railway ballast just over the approaching turnstile.

I'm immediately worried. There's nothing in the way of a level crossing.

Once you're past the gate of a foot crossing, you need to make your way to the other side as quickly as possible.

There's no dilly-dallying, or trainspotting, for that matter. As soon as you step on to Network Rail property you need to get off it. Actually, you generally shouldn't be on it in the first place, unless it's a foot crossing like this one.

Some of the foot crossing gates I've seen can house amblers or trainspotters at a minimum of 10 feet away from the rail-head, which is as close to the action as you can get while being safely shielded behind a fence. In this case, stepping over the turnstile feels exactly like trespassing as the only thing in the way of foot crossing infrastructure is a bold red sign standing solemnly, blaring, 'STOP. Look. Listen. Beware of the trains.'

Ryan isn't happy with this at all. 'No, no, no, this doesn't feel right, Francis. We'll get in trouble being here.'

'You're right, Ryan. Let's check out the other side anyway.'

I look at Traksy, a live train tracking app, to see if there are any headcodes, or trains, moving through the signal blocks between Kemble and Stroud. Nothing. I stop, look, listen, and cross the tracks. I'm disappointed. The other side isn't any better.

'There's another crossing further down anyway, Ryan, let's have a look,' I shout across the tracks.

We make our way back on to the road and, using Google Earth, I can see the next crossing is more significant: a farmer's crossing. They are a lot wider and allow farm vehicles to cross. The only issue is, sometimes these crossings are restricted for farmers' use only and are no place for a train-spotter. The law is predictable when it comes to trespassing on the railway, but farmers are not predictable at all. If you're on a farmer's crossing, you might as well be a pesky pheasant ready to be filled up with lead shot.

The muddy chevron pattern of a tractor's rear tyres leads

down to the farmer's crossing. We are relieved to see a mossy green sign indicating a public footpath next to it.

'Yes!' Ryan and I exclaim in unison.

'This crossing looks new, don't it?' Ryan points out. He's right. The moulded rubber levelling between the railheads and the road looks satisfyingly crisp. Usually the rubber is perished and crumbles under friction from a tractor or farm vehicle tyre. The gates are quite bright too, even though they are made of dull aluminium. There's also a shiny pedestrian and farm vehicle traffic light. The forest green and red lights look like sweeties and are clearly recently out of the box. They are there to tell you to stop when there's a train coming. Typically, it will give around 45 seconds' warning from my experience with other crossings, but it's usually dependent on the maximum line speed. However, as they are sensor-based, the time it takes to arrive may vary due to the speed of the train.

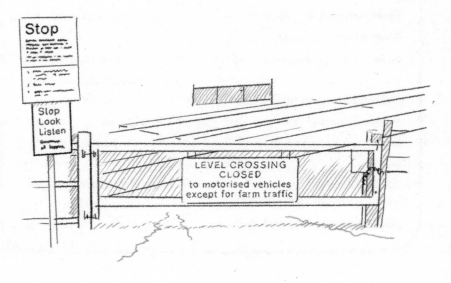

A Suzuki Jimny is parked facing away from the crossing. It seems to have just crossed, but no one is in it.

The trackside electrical boxes obscure most of our view, so we cross over. This side is better; there's less foliage here, in the direction towards Kemble. The position of the crossing itself is at the centre point of an S shape in the track, so it's the ideal spot to see around the apex of one side and, due to the lack of vegetation, the apex of the other corner too.

I find a retaining wall next to the path leading up to the gate that is a perfect place to stand on while holding on to the level crossing sign. It takes a bit of scrambling and muscle tremors in my left quad to heave myself up but looking at the view, I'm quite pleased with it. Ryan is too scared to jump up

Class 37

Build date 1960–1965

Total produced 309

Number in service/preserved 66 in service, 35 preserved

Prime mover English Electric 12CSVT (37/9: Mirrlees Blackstone MB275Tt, Ruston RK270Tt)

Power output 1,750bhp

Maximum speed 80mph

Current operators Colas Rail, Direct Rail Services, West Coast Railway, Europhoenix, Rail Operations Group, Locomotive Services Limited

Nicknames Tractor, Growler, Slugs

with me as he doesn't usually favour himself when it comes to technical climbs, even though I give him lots of encouragement and always offer a hand.

I check my app again and I see that there is a CrossCountry class 221 Voyager approaching Kemble Station. Once it has passed Kemble, it should arrive here in a matter of minutes. It's the perfect test to see if the location is suitable for watching the class 37s.

A blaring siren comes from the pedestrian and farm vehicle traffic light. This is not good. The sound of the thrashing of the class 37s will be ruined by this noise. I look at Ryan and we both shake our heads in disappointment. The class 221 whirs by, its engines working at around mid-rpm to overcome the incline, but we can barely hear them.

A smiley lady walks up the path towards the crossing, carrying a bin bag filled with garden waste. 'Hello, lads,' she calls out. 'Anything good coming through?'

'Some class 37s, but nothing special really,' I reply. 'Well, for *us* they're special, but nothing like a steam train or anything!' From my experience, when someone asks if anything good is coming through they'll stick around for a steam train but if it's anything else, they'll move on.

The lady continues to smile, and she seems interested. 'Well, take care.'

'You'll certainly hear them, that's for sure,' says Ryan as the lady looks both ways and crosses the tracks. She laughs and opens the gate on the other side with her free hand. I realize the Suzuki Jimny parked on the other side is hers. Perhaps she forgot something after she crossed – surely she can't be loading her car one bag at a time, especially if it involves stopping, looking and listening each time.

It suddenly occurs to me that she'd be a great person to

ask about our viewing spot. I shout across to her. 'Excuse me, do you know if there are any more foot crossings up ahead?'

The lady draws her head back from the passenger door. 'There's one further up the line but it's a little tricky to get to … it may even be closed now.'

'What's the best way to get to it?'

'Follow that path down there' – she points behind me – 'but make sure you don't go through the first gate on the right: I have guard dogs in there and they won't hold back,' she says jokily. 'There will be a small turnstile at the bottom of the hill. Go through it, and I think it should be easy to get to from there.'

'Thanks!' I call across.

She smiles and waves as she opens the driver's-side door.

We take the path, which soon becomes boggy and uneven. I am so relieved to be wearing my sturdy boots instead of my loafers. Ryan is sliding around all over the place and his shoes are soaked through and plastered in thick mud. I find this sort of thing hilarious. When I think of the time I've laughed the hardest, it was on a family walk across the Devon cliffs in 2010, when I was ten years old. Suddenly there was a massive downpour as we started to descend to Branscombe beach. My cousins and I were sliding on our bums down the hill, totally drenched and nearly wetting ourselves with laughter.

I can tell that walking through the mud is becoming less hilarious for Ryan the longer we're slithering along this path, but I'm still enjoying navigating the terrain, feeling like I'm in an off-road vehicle and I have to pick the best route in order to keep up my momentum and not get bogged down.

The footpath runs next to a hedgerow on one side, and along the edge of a cambered field on the other. At the bottom of the field, which probably has an elevation change of about 15 metres, is a stream. We pass some sheep who gaze at us with their almost rectangular compressed pupils. Some of them have curly horns and they definitely seem more aggressive than the non-horned sheep, who shuffle away as we come close to them. The horned sheep stand firmly in defiance.

We come to the gate the lady mentioned, on the right. A very clear 'Guard Dogs, Beware' sign in the middle.

Two Alsatians bomb towards us. My heart thuds and instinctively I step back. The gate does its job and the dogs snap at us from behind it. Their barks are loud. I'm fine with dogs in general but I much prefer cats: they're a lot easier to read and it doesn't hurt too much if they nip you.

Britain's Biggest Station

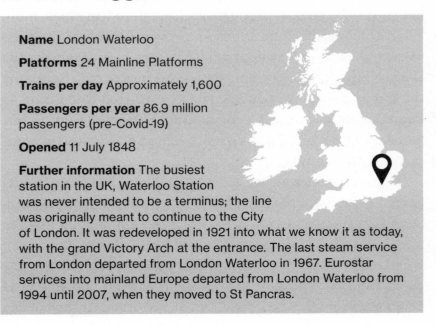

Name London Waterloo

Platforms 24 Mainline Platforms

Trains per day Approximately 1,600

Passengers per year 86.9 million passengers (pre-Covid-19)

Opened 11 July 1848

Further information The busiest station in the UK, Waterloo Station was never intended to be a terminus; the line was originally meant to continue to the City of London. It was redeveloped in 1921 into what we know it as today, with the grand Victory Arch at the entrance. The last steam service from London departed from London Waterloo in 1967. Eurostar services into mainland Europe departed from London Waterloo from 1994 until 2007, when they moved to St Pancras.

Ryan is talking about the approaching class 37s. 'They've just gone through Lydney, Francis.'

We climb up a very steep and slippery hill after going through the turnstile. There's a lot of tree cover here and the autumn leaves have been on the ground for a couple of months, turning pasty and becoming one with the mud. Ryan is having a hard time with traction. I am too, but my boots are doing a decent enough job. I advise Ryan to use the saplings to help him get up the hill and to walk up with sideways feet so that they have more of a profile against the direction of the slope.

Another gate, yellow and weathered, rises from the crest of the hill.

'Here it is, Ryan!'

Location: Sapperton Tunnel, 30 December 2021, 09:38

The view from the foot crossing is perfect. To the right the track snakes downhill, following the profile of the valley. To the left, the track is swallowed up by a dark mouth, brimmed with geometric limestone: Sapperton Tunnel.

'That looks epic up there.' Ryan's squinting, focusing his eyes on the top of the tunnel opening.

He is right. There does seem to be a good viewing spot there, but from my experience, this often gives false hope. I imagine there's a Network Rail-enforced three-spike fence, keeping the area out of reach.

We only have fifty-five minutes until the class 37s pass. But it's worth a shot.

Following what seem to be spirit lines – where deer and

other mammals traverse the same route, leaving faint distur-
bances in the leaves underfoot – we start to climb up towards
Sapperton Tunnel. I feel that I have a good sense of direction
and so I set the red dial on the compass in my brain, fixed on
the top of the tunnel.

We have to follow the tracks but the foliage and the perimeter fence barge outwards, compromising our route. There's
a small stream, deep enough to warrant a jump to the other
side. I leap across. The leaves on the other bank slide into
the water under my feet, revealing a slab of wet clay earth.
My feet treadmill wildly as I try to accelerate upwards but
there's no chance of that. I slide back into the water. Ryan's
high-pitched cackle makes it all worth it.

We press on, sacrificing our dry feet, and as we near the
top of the hill, brambles scratch our bodies. We pick them
out of the way with our fingers, clasping the rigid stems,
avoiding the thorns.

A clearing, at last! I break through on to a very rough
concrete path. Ryan is behind me, still caught in the spiky
tendrils. In a moment of rage he powers through the remaining brambles, a thorn catching his eyebrow, taking hold as he
pulls part of the undergrowth on to the path with him.

Following the choppy path, we realize there is a fence
withholding the view from us. Three metal wires are staggered one above the other, 20 centimetres apart. I can just
about see glimmers of the track through the thick bushes.

'We'll have to go back to the crossing, Ryan. This won't be
good enough.'

'I can't do that again, Francis. That was awful.' Ryan
walks on ahead in protest, shaking his head grumpily.

That's the thing with trainspotting: it's so unpredictable, even as planned and organized as it may seem. There's

always a chance a train will be cancelled, a new spot won't be good, or a farmer might shout at us. It can be deflating but I've found there always has to be contrast, a thrill of the chase, to be able to truly appreciate the highs that trainspotting has to offer.

'Oh my God – here, Francis!' Ryan is jumping up and down.

I run up to him. The buddleia and thick brambles gradually thin out to reveal a gap in the bushes, leading down to a clearing. A green strip. A stage!

We walk down and there it is – the line leading up the incline to Sapperton Tunnel, disappearing below our feet into the tunnel that we have now conquered. The trails of abraded steel emerge in the distance, hugging the first bundle of trees and swooping around along a retaining wall, past the foot crossing and cutting into the rock face beneath us. The swirling clouds just touch the tops of the trees in the middle distance. Some of the moisture is caught and held in pockets of fluff along the edges of the valley. The light is slightly dim due to the thick cloud but our excitement for this new-found spot makes the view seem totally scintillating. Keeping to the right of the clearing is our best bet as it puts us at a slight angle, relative to the track. Finding the best viewing angle is always important as it captures a nice proportion of the side of the consist – the term for a collection of rail vehicles – but it is particularly vital for this excursion in order to highlight the unusual coming together of different rail freight operating companies.

The valley starts to hum. Ryan looks at me with an inquisitive look on his face. We both press record on our cameras. We know what's coming.

The hum evolves very slowly as the two 37s power their

way up the beginning of the incline, around 2.5 kilometres away. The valley holds the sound of the English Electric engines thrashing. After reflecting and bouncing off the cushioned trees, it's soft by the time it reaches us. A Neanderthal would class this as a rather large, hungry, angry predator but at an unthreatening distance. Far enough away to continue life as normal. Not for me and Ryan though. We look at each other, mouths agape.

The noise of the exhaust grows in waves. Reflections of the hum construct and deconstruct one another, causing peaks and troughs in the amplitude of the sound, resulting in the occasional muted roar. The layout of the valley makes these moments of sonic interplay completely unpredictable and every time they arise, Ryan and I both gasp. The undertone of the exhaust note grows. Now with the waves of volume, the undertone is accompanied by a bolshy, rattly overtone that cracks the air with each cycle of the engine. The two work harmoniously and in rhythm with the deep rumbles, swaying back and forth.

'There!'

The yellow front of 37116 emerges from around the corner. Its three high-intensity headlights pan around the bend as if it's searching for any trackside train enthusiasts. The centre of the locomotive passes the obscuring tree line, which is now no longer interrupting the sound. It crashes over me, vibrating in the air like a mechanical warrior beating the timpani of trainspotting. It's working hard. This is pure thrash.

The sound is directed out of the vertical exhaust pipe and straight up into the air, where it immediately diffracts. As the train draws closer the angle of my observation increases, the sonic experience getting progressively more intense.

37116 passes the foot crossing and the final component

of the class 37's orchestra of thrash comes to play: the turbo whistle. It screams from within the engine room, tens of thousands of revolutions per minute, sucking air into its turbine blades and then tearing it apart, driven by the pressure of the exhaust. It is complete chaos, all contained within the pipework of the engine until it is blurted out from the top of the locomotive, into the atmosphere and into my body.

My heart is thumping. My breathing is fast and rhythmic. I am overwhelmed. I put both my arms up in the air as 37116 screams in my face. It enters the tunnel beneath us and shakes the ground. The echoes of the tunnel project away, back into the valley. 37510 idles behind the barrier coaches and the class 360 unit, trundling along into the ground under my feet. I stare at the horizon, listening to what's left of the reverberations in the valley. Pins and needles tingle in my hands and feet from hyperventilation. What a thrill!

Chapter Two →

The
Screaming
Valentas

Location: Willesden Junction, July 2000–August 2006

Trains have been a part of my life for as long as I can remember, certainly from the moment I could stand on my parents' knees and peer out of the windows of the local Silverlink service or Bakerloo Line. I can't have been much older than one.

I was born and spent my early childhood in Harlesden, west London, where my family lived from 2000 to 2006. My environment was busy, multicultural, and exciting. Exciting particularly because it was busy. On Harlesden High Street, seemingly always in a state of gridlock as traffic merged and collided with vehicles coming from Park Parade, life played out to the near-constant beeping and honking of horns. It was a vivid, energetic place. The cafés, barber shops and grocers were all so brightly decorated and had their own individual smells. Holding Mum's or Dad's hand, or riding on the boogie board attached to the back of my younger brother's pram while my parents pushed us, was my version of switching on the TV. I'd watch the world, taking it all in, listening to the sounds and snatching snippets of conversations, in awe at the buzz of the city and the sensory overload it delivered. The traffic in particular fascinated me, and I'd look at the car manufacturers' logos, making a special effort to remember them. Coming home, I'd stand in the bay window and line up all my Hot Wheels cars on the window sill to admire them. I spent a lot of time at that window. The street outside was quite narrow, so I'd often see two drivers meeting head-on, with one of them having to reverse to allow the other to pass. This is normal, but in Harlesden the manoeuvre would usually be carried out at around 30mph, engines red-lining while music blared out from the speakers. My eyes would be on stalks.

A Mitsubishi Galant Estate used to park a few cars to the left of the house on the opposite side of the road. It had a very aggressive angular body kit, a frowning face but a more jovial rear end. Looking back, it could've been a VR4, but I suspect the owner may have been wanting to impersonate one by adding jutting panels, bumpers and spoilers to a cheaper model. The best thing about the car was the exhaust pipe. Only one. It was absolutely huge and would fart profusely in the morning. I would run to the window to see it. I was obsessed with exhaust pipes, convinced that the exhaust from the pipe itself was what propelled cars forward. I held this belief until the age of seven. Until then I would routinely check out the exhaust pipes of cars, I think because I also recognized the correlation between a fast car and the type and number of exhaust pipes it had, and ultimately the degree of noise it made. Visits to see Granny in Notting Hill became the Mecca of my car-spotting days, *circa* 2007 to 2011, when I used to run around with a pad and pen, taking note of every single supercar or sports car I could find. At one point I refused to take down any more Porsches because they

were so common in her area. We only had the farty exhaust pipe cars in Harlesden.

My parents didn't own a car until 2003 so until then we used to get the train everywhere. Willesden Junction was the nearest station, and we were very lucky that it was too: 72 stock Bakerloo Line trains rattling along on the low-level platforms, cam shafts tapping away; class 313s, operated under Silverlink, serving the high-level platforms, EE GEC traction motors groaning while the compressors rattled away; and West Coast Mainline carried Virgin Pendolinos at

Class 08

Build date 1952–1962

Total produced 996

Number in service/preserved 82 preserved, 10 converted to class 09s, 6 converted to class 13s, 5 exported to Liberia, 100 in service, remainder scrapped

Prime mover English Electric 6KT

Power output 350hp

Maximum speed 20mph

Current operators British Railways, InterCity, Network SouthEast, Rail Express Systems, Freightliner, Eurostar, DB Cargo UK, GNER, National Express East Coast, East Coast, Virgin Trains East Coast, LNER, Midland Mainline, East Midlands Trains, East Midlands Railway, Arriva Rail North, Northern Trains, Harry Needle Railroad Company, Foster Yeoman, Mendip Rail

Nicknames Gronk

around 70mph, Silverlink class 321s, Virgin Supervoyagers, HSTs, class 86s, 87s. Not only that but there was a scrapyard on the other side of the tracks, mechanical claws swinging around, conveyor belts flinging out shredded metal into piles. There was once a fire there too, a huge one, after some petrol left in a fuel tank caught alight. For a boy obsessed with things that moved and made a noise, I was in heaven.

The journey to school was fantastic as we'd pass Willesden Traction Maintenance Depot (TMD) on the Bakerloo Line, which was always a hive of activity. The 08 shunter was a particular favourite, in its woodlouse-like form – small, compact, robust, and always moving slowly. Affectionately known as 'Gronks', supposedly after the sound they make, unusual yellow and black chevrons run top to bottom on their rear and the wheels are interconnected by a rod, similar to an old 0-6-0PT (pannier tank) for example. Each day something would be different at the TMD; that's what I kept a look-out for and looked forward to. Retrospectively, I feel like this ignited the love I have for the trainspotting aspect of my hobby. Continuing along on the Bakerloo Line, I used to get off at Kensal Rise, where the 72 stock tube train would, without fail, perform an electrical discharge every time it left the station heading west into the tunnel.

The playground at school had swings, climbing frames, hula hoops and sand boxes, and it was almost too perfect for a young train enthusiast. Fortunately, the North London Line hugged the edge of the school fence, close to where a climbing frame led up to a wendy house which you could then slide down from. I used to stand at the top of the wendy house looking over the fence and listening. It was my own little look-out, with the occasional deviation to join a game of tag or to talk to a friend. It was a treat if a freight train

passed during break time. EWS and Freightliner class 66s were most common; the maroon and yellow and the green and yellow livery respectively were particular favourites. I even remember a dream I had about a Freightliner class 66 on the other side of the fence. It is dark, and I'm in the play-ground at night with no one around. A Freightliner class 66 stops at a fictional signal right next to the playground. A skeleton is driving it. He looks at me terrifyingly, and as he starts to pull away from the signal, embers shoot up from the top of the exhaust, thousands and thousands of embers. That's when the dream stops. I've no idea what it meant.

Once at home, I'd play with my Hot Wheels or my Brio train set. Brio, a blank canvas for a young railway engineer, confines creativity to the radii of the curves and the lengths of the straights, which, in reality, is the best introduction to creative building as it's so simple to grasp and easy to put together. My track layouts would only last a day before being pulled apart for a new start, but I loved creating them. I loved drawing road maps too, such as the route from our house in Harlesden to Granny's house in Notting Hill, trying to accurately recreate the journey from memory.

Granny would pick me up from school every Monday. I'd go back to her house and have strawberries dipped in sugar, and peach yoghurt. Under the stairs at her house was a cup-board. In it was a Henry hoover, a mop and bucket and my Hornby train set – my first proper train set. Granny had very kindly bought it for me in the summer of 2004 when we visited Pecorama in Devon, home to the model railway company PECO's headquarters, which had a fantastic model railway exhibition and a narrow-gauge steam railway. After looking at the exhibition, I imagine it was quite evident that I wanted to get stuck in. When we left, Granny surprised me with a GWR steam freight set with a 0-6-0 pannier tank, a few freight wagons and a brake van. Being four at the time, my curious, clumsy hands would've made mincemeat of the delicate models had they been left in my possession so Granny very wisely kept the set at her house. The laminate flooring in her kitchen was the perfect surface for a model railway – no fluffy carpets to get stuck in the components.

We'd take it out of the cupboard. The polystyrene casing would slide out from the cardboard sheath. Track, controller and plug would be in one section, the locomotive and its carriages would sit in the other, in perfect extruded cuts of their side profiles. Wiggling them out was a little bit tricky. There were small indentations in the polystyrene to allow fingers to clasp the desired component of the train set, but a bit of the polystyrene would always crumble and fall on the floor. Once the track was out, aligning fishplates on both sides of the track would be next – a near-impossible task when I started but one that soon became second nature. Lying down on the floor, I would carefully examine the alignment of the wheels with the track, sometimes feeling the vibration when the wheels rolled over the plastic sleepers, lifting that particular wheelset up until it slotted on to the gauge of the railway. It was the most satisfying feeling, re-railing a derailed carriage – rough, rough, rough, then smooth and frictionless. I would be happy for hours watching the train go around, sometimes modulating the speed, changing the points by hand and reversing the freight consist into the siding. It made me the train driver and the observer. It was my own little world; even though it only featured tracks and a train, it was all I needed to make me happy.

Granny once invited a friend of hers round when I was playing with my train set. I heard them say hello at the door and then heard her big boots clonking on the floor. I remained lying down as she walked into the kitchen, eyes fixed on my 0-6-0 pannier tank, whizzing around at full speed. Whoosh! A boot the size of a small housing estate swung into my zone and connected with the side of my pannier tank; it departed my world and crashed across the floor of the kitchen. I scrambled after it. Its moulded plastic shell had completely detached from the chassis. I'd never seen its insides before, its wound copper motor and faceless front. I was quite upset. The shell went back on but one of the clips had snapped, making it prone to rapid body detachment. Granny's friend was very apologetic. I was a lot more wary of running my trains on her floor after that, and later opted for an upgrade to her dining room table.

One of my trainspotting highlights from those early years was a moment I could not have predicted at the time. My parents often took me to the station to see the trains. I was perfectly content with spotting my Silverlink class 313s and often drew them, sometimes asking my dad to draw them if I wanted something a bit more realistic. What I'd never expected to see was one of my poster trains, a class 373 Eurostar, the same train I got a model of for Christmas in 2005. I was in the car with my family, Dad at the wheel of his silver Rover 25 manual with a partial leather interior. We were driving along Scrubs Lane, heading back towards Harlesden, when, as I stared out of the window from my booster seat, I saw the mechanical harp that runs along the side of a class 373 Eurostar and the handsome, distinctive chin that pierces the air at 300km/h from London through to Europe. It was just sitting there next to a Big Yellow Storage. It was like

Britain's Smallest Station

Name Berney Arms

Platforms 1

Trains per day Request stop

Passengers per year 348 (2020/2021)

Opened 1 May 1844

Length of platform(s) 18 metres

Further information Also the quietest station

meeting a film star in the back of Iceland. I can remember being excited and confused all at once. I was insisting to Dad that we needed to turn around so I could go and have a look. We dropped Mum and my brother back at home, then Dad and I headed back out. Crossing fingers was a recent discovery for me and I was crossing as much as I could, hoping it would still be there. We pulled into the car park and it was there, just as before. We got out and walked up to the steel three-spike fence, which felt 10 metres tall. Behind it, the 373 stood on the tracks. I had held this exact class of locomotive every Monday and watched it go around and around on the loop of my world, now it was here for real and the tables had turned. It was quite intimidating standing next to it without the added height of a platform. Peering to the right I could see the trailing coaches.

Now that I know a lot more about the rail networks, I know why it was there. The Eurostar used to run from London Waterloo, shifting above and over the domestic line into Waterloo Station via International Junction, through Brixton and out to Ashford in Kent, then to the coast the twentieth-century way, taking power from the 750V DC third-rail system. They would be tucked into bed at the North Pole depot at Wormwood Scrubs, just next to the Great Western Mainline. To access the depot they'd travel along the West London Line, heading through Kensington Olympia and Shepherd's Bush then diverging just before the bridge over the top of the Great Western Mainline, and descending next to the line into the £76 million depot. The siding next to the Big Yellow Storage was an infrequently used single track. If I had known about the North Pole depot, which is so close to Harlesden, I would have been asking my parents to take me there every day.

I find the original Eurostar infrastructure fascinating as a lot of the lines used weren't bespoke, which would lead to scenes in Kensington Olympia, for example, where a sleek 300km/h class 373 would be passing a dusty quarry train or scooting past a cramped networker on the way out of the city. Regional Eurostars were considered once. They even built a depot in Manchester to accommodate a fleet of class 373s. The idea was that passengers would be able to get on 373s at Manchester, Glasgow, Cardiff and Plymouth (the latter two being subsidized by stock used on the Great Western Mainline) and then travel directly to Paris. However, due to competition from cheap air travel, these plans were scrapped, even though signage was installed in the planned terminuses further up north. Some of the 373 sets were leased to Great North Eastern Railway (GNER) for a period up until 2005, which made things very exciting on the East Coast Mainline, particularly as they wore GNER's stealthy black and red livery. In 2007, the terminus for Eurostars in the UK changed to St Pancras, in north London, and HS1 was created. This enabled Eurostars to depart from the station and immediately dive right over the East Coast Mainline then head underground, remaining there for 19 kilometres, emerging at Dagenham and continuing to Ebbsfleet and Ashford International. From the moment the class 373s commenced operation out of London St Pancras I didn't see them again until travelling from St Pancras to get to university in Nottingham eleven years later. Sadly, by this point, class 374s had barged most of the class 373s out of the way, new, faster units that in my opinion look a lot uglier and don't make the amazing sound class 373s make when they pass by at maximum line speed. The gaps between the 373 carriages create pressure differences that emit a sound every time each of those gaps passes the observer. *U-U-U-U-U-U-U-U* ... nineteen times.

When I was six, my parents decided to move to Somerset. The options for secondary schools weren't looking good for me and the pollution of London was not good for my brother's asthma, so it seemed like the right time to up sticks. I went house hunting with Mum, catching the train from London Paddington. Standing in the main concourse, the arches above seemed to touch the sky. It was very loud, the station full of ground-rumbling bass and rattly overtones. A lot more dirty than it is now too. Dark purple First Great Western HSTs sat in every bay, angular noses all lined up like punters in a tavern grumbling for their next ale. A pink stripe ran from the front of the powercars, rising and fading off towards the rear. A gold stripe started halfway down the powercar, crossed all the coaches, only interrupted by the white doors, and finished on the rear powercar, the whole stripe highlighted by an accent of pink running along the top. The sound attracted me to them: it was like those farty exhaust pipe cars but way louder and deeper. The yellow fronts gave a sense of familiarity too. A massive plume of smoke poured out from the top of one of the powercars on platform 1, rising up to the top of the canopy, dulling the light shining through. It wasn't moving but I could hear a growl growing. Then, something started to pierce the bass tones like an angle grinder. A high-pitched noise like I'd never heard before. *WUUUUUUUUU!* The turbos on either cylinder bank of the Paxman engine had started to spool up, creating a deafening whine. At this point the train was moving slowly but the pure thrash from the engine gave the impression of such speed, power and energy. The whine kept growing, clag being projected violently into the ceiling. It was screaming now. *EEEEEEEE!* I put my hands over my ears and looked at Mum, who didn't seem to be as engaged in its departure, then looked back down platform 1, the powercar just disappearing out of sight behind the set on platform 2. The driver had notched back the power a bit as the trailing powercar left the canopy, suddenly reducing the decibel level in the station hall. I felt so jumpy, fascinated, engaged, just the same as I do now when I experience the pure thrash of a locomotive. It dawned on me that we'd be riding on one of the class 43 HSTs to Somerset – the first journey on one for me, the first time I rode on what is now my favourite train.

The class 43s were certainly in their prime when I had my first trip on one in 2006. They used Paxman Valenta engines, nicknamed Screaming Valentas. Two 2,250hp engines on either end, mostly quiet in the centre carriages, just the operational noise of the wheels; towards the powercars though, the constant presence of a monster just over your shoulders. Paxman Valentas were removed from all HSTs by the end of 2010 in favour of the more reliable, quieter, less polluting MTU 4000 series. These engines run in all mainline HSTs that are still operational – no screaming, just rumbling. I now watch class 43 HST Paxman Valenta videos on YouTube as a form of relaxation, remembering the few memories I have of them as a child.

Location: Bruton, 2007

My cousins lived in Bruton in Somerset, on the Wilts, Somerset and Weymouth Line. The line passed through the centre of the village, running close to the lowest point, along the rough guide of the river Bru, trains blasting through at 90mph. Rising from the Bru and the train line, houses and schools sit on the sides of the gently undulating hills. The tops of the hills are still green with rough hedgerows and tree lines dividing up the farmers' land. My cousins' house was on a narrow street, one of the more elevated terraces. You could hear when an HST was coming around thirty seconds before it passed through, which at a speed of 90mph is a long distance. The rumble would come first, then the scream and the whooshing of coaches, followed by another scream and a rumble. I couldn't see the HST standing on my cousins' hill but the sounds were more than enough to stimulate me, especially with the echoing and reverberations around the higgledy-piggledy rooftops.

Important Railway History

1812
The first industrial use of an effective railway. The Middleton Railway in Leeds introduced a steam-powered rack-and-pinion locomotive that replaced horse-drawn coal wagons

15 September 1830
The opening of the Liverpool–Manchester Railway. Utilizing 1,435mm Standard Gauge, the first intercity railway proved to be a massive success and kickstarted the development of further railways

15 August 1846
The Regulating the Gauge of Railways Act 1846 made any new railways Standard Gauge (unconnected to the then current Broad Gauge network)

10 January 1863
The first underground railway. Running from Paddington to Farringdon in cut-and-cover method tunnels. Hauled by steam locomotives and gas-lit carriages

23 May 1892
The last Broad Gauge line between Exeter and Truro is replaced with Standard Gauge, making the whole network Standard Gauge

3 July 1938
The A4 *Mallard* achieved the steam speed record of 126mph, just south of Grantham in Lincolnshire

1 January 1948
Formation of British Railways, nationalizing the railway

1955
The Modernisation Plan of 1955. Steam locomotives were ordered to be banished from the railways and were replaced by fledgling diesel and electric locomotives. Standardization increased, as did the overall efficiency, reliability and speed

1963
The Beeching Report. With British Railways losing money, Dr Richard Beeching was tasked with returning the railways to profitability and thus closed one-third of the stations on the network

April 1965
British Railways relaunched its corporate identity as British Rail, introducing the blocky, clear, double-arrow logo that we know so well today under the National Rail brand

1 November 1987
Powercars 43102 and 43159 broke the world speed record for a diesel train, achieving a speed of 148mph

6 May 1994
The Channel Tunnel opened, enabling fast access to the continent with services running from London Waterloo

14 November 2007
HS1 opened, offering high-speed links into Kent at speeds of 140mph and allowing Eurostar services to travel at 186mph in sections in the UK

24 May 2022
London's new subterranean Elizabeth Line opened, enabling quick services from Paddington to Abbey Wood. Later to be extended to allow through-services from Reading

Waiting for the train at Bruton Station to go back to our town in Somerset, Frome, would sometimes take close to an hour. Only services to and from Weymouth and services to Yeovil would stop at Bruton, all others would blast through. I was waiting with my mum on platform 1 when suddenly a class 43 HST passed through at line speed, screaming its head off. I was standing in front of a sign, my head resting against it, the bass and screaming reflecting off the sign, amplifying the chaos in my ears. I jumped out of my skin, terrified at the noise, my eyes wide open watching the blur of purple and pink. The coaches whooshed by, slightly more soothingly than the aggressive Paxman attack, followed by the next powercar, more predictably thrashing the same way but still intensely scary, its red lights blaring off into the distance, the top of the locomotive distorting light passing through due to the heat from the exhaust. I burst into tears as it was such a shock, and Mum gave me a reassuring hug. I remember exactly how I felt then; if it was now, the same shock and sensory awe would send me into an excited craze.

I miss the Paxman Valenta. Those two memories, at Paddington and Bruton, are the only ones I have of the engine in operation. It is one of my goals in life to try and get one to run once more. It would be my dream one day to reinstall a Paxman Valenta in a class 43 HST. I need to feel it again.

Moving away from London was a big shock to the system. The vibrancy of Harlesden High Street, the multiculturalism, the density and the chaos was replaced by open fields, flowing roads and sparse train lines. I was quite content though. I made some friends and we had a house with two floors. My model railway was able to grow in the conservatory, expanding with houses, stations and people. After coming back from school I could dive in, build and create. I was happy, but I left things behind in London that I missed: Willesden Junction's high-level platforms, the Pendolinos blasting by at 70mph, the scrapyard and, most of all, my purple, green, yellow and white Silverlink class 313s.

Chapter Three →

Dick Mabbutt

Location: Barnham Station, 20 September 2021, 15:41

My 2003 Nissan Micra bounces as I pull her into Barnham Station car park. I call her Lucy. She was my first car. Six months ago, she was retired to the driveway of my parents' house where she stagnated and started to grow moss on her rubber sills. My second car, no name yet, is in for a service: I suspect there's a timing chain issue, and the front right wheel needs balancing. Lucy is glad to be back on the road, although she doesn't know it will be transitory. I have a strong relationship with her, lots of happy memories, and I know her quirks – the little squeak her automatic gearbox makes as she shifts from second to third, just to mention one of them. She feels like a go-kart at low speeds but has diabolical acceleration above 40mph. My new car doesn't have many issues in the acceleration department, it's a 2007 Mercedes E-class, though I have not yet connected with her much.

I have around twenty-five minutes until my train to London Victoria. It's fantastic weather for October: the shallow light of winter hasn't touched today, the sky is welcoming, and the clouds are wispy and sparse. There's a parking space right at the back. Usually I reverse into these bays; at work it's mandatory as it vastly improves the flow of traffic when everyone finishes their shift at the same time. I pull Lucy's nose straight into the space, ignoring my usual parking routine due to a class 313 PEP just pulling into platform 1. I'd quite like to watch it while I gather my thoughts and belongings. The wire link fence separating the car park and the tracks, intertwined with some kind of climbing plant, reveals just enough of the Southern class 313. They mean a lot to me as the units served my old station, Willesden Junction, under the train operating company Silverlink. I remember their ancient traction motors groaning as they pulled into the platform. The doors hissed as they opened. The compressors droned and clicked as they took on

Class 313

Build date 1976–1977

Total produced 64

Number in service/preserved 20, 44 scrapped

Prime mover Traction motors: English Electric GEC G130AZ

Power output 880hp

Maximum speed 75mph

Current operators Southern, Network Rail

Nicknames None (I would call them groaners)

passengers. The traction motors then twisted the steel axles of the motor coaches with maximum tractive effort, groaning inversely to the way they groaned on the way in, as the pitch of the sound of the traction motors is proportional to the speed. *Eeeaaaooowww* on the way in. *Owwwaaaaeee* on the way out. These three-car units are happy to be singing at speed but they are a little bit grumpy as they come in and out of a station, that grumpiness being particularly obvious when the carriage has vibrating temper tantrums as the traction motors operate at lower speeds, hitting the resonant frequency of the carriage shell, vibrating everything, including passengers, at that frequency. The Silverlink franchise ended in 2007 and the class 313s were transferred to London Overground. Then, once the class 378s barged them out of the way with their superior standing capacity, they were transferred to Southern, where they now potter around along the south coast. Their current livery is white and mint green with a typical yellow front. Familiar beady eyes (two on the left, three on the right) look at me in the same way they used to on Willesden Junction platform 5.

Here at Barnham, a gentle slope leads from an underground passageway up to platform 3. The station, just on the outskirts of the village, backs on to the car park on one side and on the other faces flat fields and rustling trees, still holding on to their leaves. The canopy on the opposite side casts shadows across the passengers waiting for a service towards Brighton or London Victoria. Most of those waiting are school children – there are huddles of them laughing and joking, some standing, others sitting with their backs against the wall. There are a few commuters but they are greatly outnumbered at this time of the day. I stroll to the western end of the platform; it's quiet here and the line is straight in both directions for miles. Half of the class 313's front carriage pokes out from behind the waiting room on the other platform. It's perfect light to get a photo of it. I feel so relaxed here. The tracks from platforms 2 and 3 individually diverge, heading either to Bognor Regis or in the direction of Portsmouth, splitting just after the western end of the station. The lines heading to Bognor Regis then join with the line from platform 1 that splits in order to marry up to them. But it doesn't stop there. Even before the line from platform 1 splits, there is a point that allows trains on platform 1 to cut across both lines from platforms 2 and 3 heading to Bognor Regis, darting straight across them and linking up to the line towards Portsmouth from platform 2 and later leading to the line that links up to platform 3. I stare at the duality of metal cutting, dividing and merging while always having a partner rail set at 1,435 millimetres. It's deeply satisfying, even more so given that there is a third rail too, following the parallel of the two rails, providing the trains with 750V DC.

I squint and stare down the tracks in the eastern direction, and spot another class 313 approaching. I can tell what

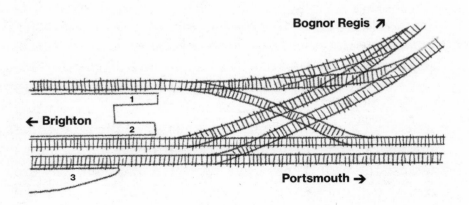

class units or locomotives are from quite a distance, even just from the headlight formation when it's dark. Two 313s is rare. To get a shot of both of them at once will be mega. The complex shade of the trees and foliage flickers across the front of the unit; occasionally direct sun beams on to its yellow front, catching a twinkle in the uneven red and clear lights. The 313 takes a little while to carry itself to platform 2. I can read its number just before it hits the shade of the canopy: 313215. The unit decelerates to the sound of the deeper, descending groan. I know it so well. I take a picture.

Gauges

The Standard Gauge width of 1,435mm was adopted after it was used in the construction of the first intercity railway in the world, the Liverpool–Manchester Line. The Grand Junction Railway and the London and Birmingham Railway followed suit. Isambard Kingdom Brunel, on the other hand, opted for a wider 2,140mm, known as Broad Gauge, for the London to Bristol line, later known as the Great Western Railway. In 1846 the Regulating the Gauge of Railways Act 1846 made any new railways Standard Gauge (unconnected to the then current Broad Gauge network). The Great Western Railway was converted to Standard Gauge in 1892 and the UK has since continued the use of Standard Gauge. It can be found across the world: in North America, the majority of western Europe, North Africa, the Middle East, and on high-speed networks in Japan and China.

Broad Gauge is now accepted to mean any track significantly wider than Standard Gauge and is used in India, Russia, the Baltic states, Georgia and Ukraine, Mongolia and Finland, Spain, Portugal, Argentina, Chile and Ireland.

Clatter clatter, CLATTER CLATTER – the clattering of wheels over points cracks over my shoulder. I turn. I gasp! Without warning a third train, 313210, trundles into platform 3. I scramble to record a video. There's no way! My eyes widen, mouth agape. I just manage to record the sloped start of the platform rise and level out alongside the unit and it passes both 313208 on platform 1 and 313215 on platform 2. Only moments after the first drone of familiarity, there it is again: *eeeaaaooowww*. I'm giggling at how I was caught so off guard. It's not such a sensory climax as a class 37 thrashing or a class 43 HST blasting by at speed, it's far quieter. It's the unexpectedness of the situation that gives me an overwhelming sense of joy, a massive burst of sheer delight that continues to reverberate in the hours that follow. I am beaming. 313210 stops almost perfectly next to 313215. I take another photo, the trio of class 313s all in one shot with 313210 and 313215's compressors firing away, 313208 watching from afar as it cools down. The school children pile on and 313215 departs from platform 2. It rockets over the points at the end of the platform, *owwwaaaaeee*, its GEC traction motors maxed out, *eeeeeee CRACK*: as the third-rail shoe disconnects from the third rail over the points, there is a massive electrical discharge, searingly bright, retina-burning energy. The traction motor re-engages and picks up on the pitch it left off on before the electrical discharge: *eeeeeee* ... The train rocks back and forth over the next set of points and the three dimpled carriages barrel off into the distance.

313210's door warning chirps away at any last-minute dashers running up the slope to platform 3. The doors simultaneously initiate the closing sequence, a pneumatic drum roll as some doors close before others, each contributing a tap on the snare, with varying intensity depending on the distance from my position. 313210 has a clear run ahead with a dead straight stretch to the next station, Ford. It floats off eastbound, traction motors fading off down the line. 313208 is still there on platform 1. It spends a little while with its doors open, inviting passengers in for the short shuttle run down to Bognor Regis. I would jump on if I had time, just for the ride.

I walk back and forth with a smile on my face. I dawdle around a lamppost, then decide to sit down. The rough tarmac is warm from the sun and I welcome it on my palms. A class 377 whines into platform 3, with less character and less soul than its companion, 313208. The 377 is the train I will be taking to London Victoria. The doors open in unison. I get on and settle down. Bing bong: 'This is a Southern service to London Victoria' the announcement crisply informs the near-empty carriage. The doors snap shut, traction motors engage, departure is swift and efficient. 313208 remains on platform 1, watching us happily.

I have to change from a Southern service to a Thameslink service at Three Bridges, an easy switch from a class 377 to a class 700, the latter being the newer of the two and a demonstration of a train that serves its purpose well: it can hold many passengers and its acceleration is more than sufficient. I haven't had much experience with 700s but their efficiency makes them a bit sanitized and bland, although their traction motors make an interesting noise. From a standstill to line speed, it sounds like a digital turkey clucking, increasing its frequency as the train speeds up until it reaches a constant whine. I am sure they will grow on me with time.

Location: Gatwick Airport Station, 20 September 2021, 17:02

'The next station is Gatwick Airport.' The light has faded, rain has started to streak across the windows, and the reflections of the interior of the train dominate my view. The train slows and we arrive at Gatwick Airport. There has been considerable engineering work here – new track, new platforms, new lifts and stairs – and I'm quite keen to check it out. When new track is laid, the smooth, clean concrete sleepers look like chocolate bars equidistantly placed; the rails are so long, the stiff metal flexes under its own weight. I'd like to see it, but I think I will stop over on my return journey.

So after a night in London, I return to Gatwick Airport. The familiar dead white light casting over the platform edges has been replaced entirely towards the centre, where passengers descend, choosing either platform 1 or platform 2 as a place for a moment of thought and consideration. Blue lamps hang from the centre of the canopy. They emit a deep electric-blue glow that reminds me of fly catcher lamps, where the light draws flies in then zaps them with electricity. The filament looks similar. I first noticed these lamps a few years ago and found out that they are used at platforms where there is a high rate of suicide. The blue is the same shade and intensity as the flashing lights of emergency vehicles which, for those who may unfortunately be questioning the purpose of their lives, will deliver the safety and support that they need. Seeing these lamps at the platform offers a subconscious reminder that help is there. I often keep an eye out for people who look particularly sad, pacing back and forth near the edges of platforms where express trains pass through. I'm not the right person to deal with a situation like that but station staff are properly trained and there is no harm in raising a concern with them.

I'm speaking with the dispatcher about the blue lamps as a Southern class 377 pulls in. I give the driver a wave and the dispatcher quickly cuts off our conversation in order to make sure everyone safely gets off and boards the train, which is bound for Brighton. I nod to him as I walk back into the foyer between the platforms, heading over to the other side to check out some of the completed engineering works. I stroll along the platform examining the virgin track, ballast and cabling. The metal clips securing the track to the sleepers are wrapped in blue polyurethane. It's a pleasing mix of textures, like a good combination of fillings for a sandwich. I can imagine the sleepers being quite brittle and crumbly like toast, the ballast being crunchy bits of fried onion, the track being chewy streaky bacon, the securing clips being robust, rubbery plum tomatoes, and the oil and grease that will eventually seep on to the tracks being some kind of condiment. Not the most conventional sandwich perhaps, but I do find my brain sometimes likens different things in a rather random way and I find it quite satisfying.

73965 *Des O'Brien* passes behind me. I scramble for my camera to take a video. 'For goodness' sake, why wasn't I checking Traksy?' I think. 'I could've prepared for it and got a good shot!' I press record just as the front locomotive burbles by. Its blue, orange and yellow livery is distinctly GB Railfreight – one of the UK's main rail freight operating companies. There are only five 73/9s in GB Railfreight's possession; they are rare and a real relic of the past. It's on a test train, the Ultrasonic Test Unit, or UTU for short. The test train carriages are bright yellow and a combination of 1960s mark 1 and 1970s mark 2 carriages retrofitted with high-tech track monitoring equipment. Through the tinted glass there are rows and rows of LCD monitors, occasionally

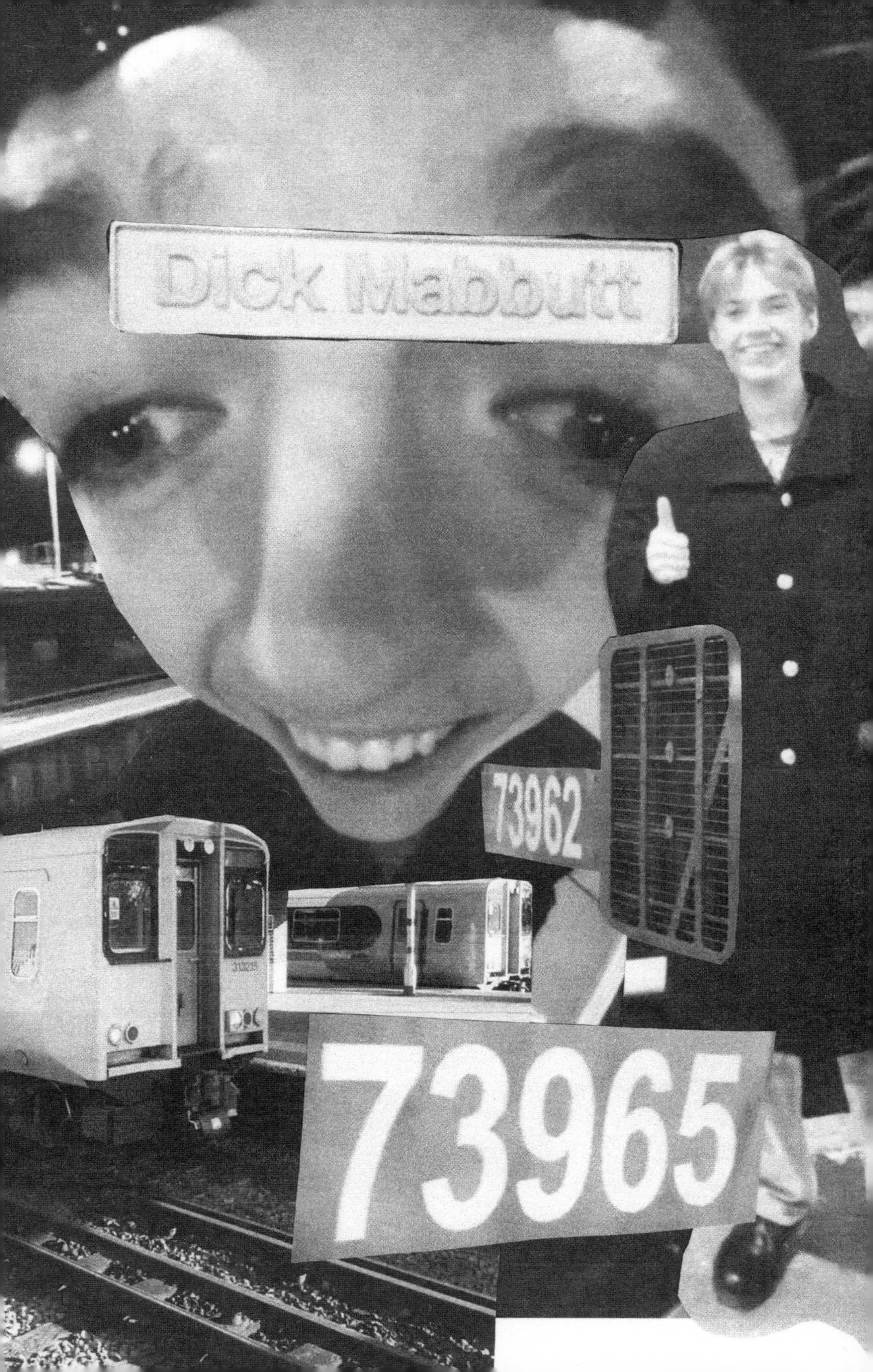
Dick Mabbutt
73962
73965

interrupted by the head of a technician. They are the eyes looking at the data, the equipment is the ears listening for any track imperfections through high-frequency reflections. It's travelling well below line speed and the MTU R43 4000 exhaust note fades down the platform.

I know that with UTU test trains there's always a second class 73/9 on the back. It suddenly dawns on me: 'Could it be 73962 *Dick Mabbutt*?' I've only ever seen this locomotive on YouTube and it has been on my trainspotting bucket list ever since. Its unfortunate comedic name and its endearing shoeboxy shape have drawn me to it. (It is named after the much-loved former chief electrical engineer of the British locomotive manufacturer Brush Traction. Richard Mabbutt, affectionately known as Dick, worked his way up from an apprenticeship and served the company for forty years, only to pass away unexpectedly in 2013. A year later, 73962 was named in his honour.)

Britain's Longest Station

Name Gloucester

Platforms 4

Opened 4 November 1840

Length of platform(s) 602 metres

Further information Gloucester has the longest single railway platform in Great Britain (although Colchester's combined platforms 3 and 4 stretch to 620 metres)

'Please be 73962, please be 73962,' I'm thinking in my head.

The platform is scattered with returning holiday-goers and a few commuters and I can see they are slightly puzzled by the bright yellow carriages. I always find it so amusing to see the reactions of non-train enthusiasts to unusual trains. The edges of the rear class 73/9 catch the light of the platform. The cast nameplate edges out slightly from the pock-marked panel of the fifty-six-year-old locomotive. It looks to be a long nameplate, certainly not 73961 *Alison*. The numbers come into view: 7 ... 3 ... 9 ... 6 ... 2. 'Oh my God, it's *Dick Mabbutt!*' I cheer out loud. The passengers on the platform turn to look at me but I don't care. I think it's clear that I'm excited by the locomotive and I haven't just blurted out a funny name. Their stares fade into the station as I pan across with my camera, the canary yellow of the rear paired with the staring red tail lights that glare at me. I cannot just let this be the extent of our encounter; I need to get a well-planned shot of it.

I check live.rail-record and see that the test train is pulling into Brighton Station at 00:15, and that it takes quite a convoluted route as it's examining the track. Between now, 20:32, and the time when the test train is due in, I can get the train back to Barnham, pick up Lucy, drive to Brighton and wait for the test train to arrive. But then I remember I have work in the morning. I have to get up at 05:30. The drive from Brighton back to where I'm staying in Portsmouth would be around an hour and twenty minutes, meaning only a few hours' sleep. What if I just sleep in my car? That's it! From Brighton, I will drive to my place of work, park up and sleep there. That'll give me another two hours of sleep as I won't need to commute in the morning. I even have my work clothes with me. I don't make a habit of sleeping in the office car park but when moments like this happen I need to take the clips off my wings and fly. With trainspotting there's always a gut instinct when it comes to choosing a location, station or just chucking in your dice. Ultimately, it's the adventure and the unpredictability that makes the hobby so addictive. Brighton Station could end up being as dead as the night, a dud, or it might just be my nirvana for a precious few minutes.

Barnham is cold. Fortunately, I've chosen to wear my 1980s British Rail trench coat to keep warm: its thick wool traps my body heat. The cars are covered in condensation. Lucy's bulbous headlights glisten as I approach. Her eyes used to be clear and white, now, due to the way the sun has degraded the plastic, they are tinted yellow. The condensation conceals this a little. I take out the key and press unlock a number of times – *click click click*. The car doesn't open. I move to the rear left of her, where her aged circuitry is usually most receptive to the signal from my key. Mercifully, her indicator lights flash in greeting.

Location: Barnham Station, 21 September 2021, 22:12

I set Google Maps to Brighton Station. Approximately one hour, 27 miles. Chasing cat's eyes most of the way, the traffic only starts to build around Brighton seafront where the bolshy teal-and-white taxis assert themselves as lanes merge. I make my way around the back of the station to the car park, plonking Lucy on the first floor in the corner. I have around an hour and ten minutes until 73962 and 73965 roll in so I wait a moment in the car. I check Traksy and see they are at Eastbourne, a terminus, where the driver will change ends and lead the rear locomotive, 73962, into Brighton, just as I had hoped. Time to head up to the station.

Brighton Station is a curved terminus with a mid-nineteenth-century canopy – absolutely stunning, refreshing, and a reflection of Brighton as a city. At night, the place is shrouded in orange light from lamps that hang directly over the platforms, highlighting the curved aerial profile of the station. It's so orange it seems to be an aesthetic choice over the cold white lights typically used in London terminuses. It's quite a coincidence, actually, because the blue of the station canopy is very similar to GB Railfreight's blue and the orange lamps replicate the highlights of orange in the livery too.

There are a few gateline staff positioned along the entry/ exit gates. There don't seem to have been any recent arrivals so they are only assisting with passengers boarding the

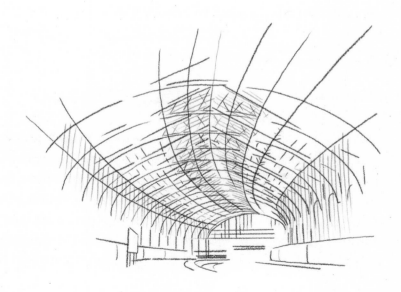

last trains of the day – considerably less busy than normal and ideal for me. I wander up to one of the staff members. I always feel a sense of anticipation when asking gateline staff if it's possible for me to go on the platform to trainspot. On every occasion I have been met with zero resistance and a smile but I'm always thinking in the back of my mind that it's an unusual thing for someone to ask.

'Excuse me, I was wondering if it'd be at all possible to let me on to the platform as there's a special test train due on platform 5 that I would like to see, please?'

They turn around and nonchalantly open the gates with their card. 'Yeah, sure.'

I nod and smile. 'Thank you!'

'You're welcome,' they reply.

I am so glad I'm wearing my thick wool trench coat, as it's freezing. I traipse up platform 7: 313201 is sitting on platform 8 waiting to take passengers back along the coastline.

It wears a wonderful heritage British Rail livery, commemorating its service in its final years. I take a close look and also examine the platform width, looking at the different shades of pavement. 'I wonder if there used to be another two lines in here that were later removed and filled in,' I think to myself. There's a Rail Head Treatment Train, or RHTT for short, on platform 3. The driver is just changing ends. It's an unusual-looking train that features two cabs on either end of the two-car unit. In between there are two rounded-off cylindrical tanks, bright blue and holding thousands of litres of water. The water is pressurized and sprayed directly at the railhead. This prevents build-up of a paste-like substance caused by wet leaves falling on the track and being compressed by the wheelsets running over it. Friction is a critical component in the dynamics of the railway: without it, trains would not be able to accelerate and, more importantly, brake. The leafy gunge vastly reduces friction and causes difficulty in accelerating and braking, impacting journey times and the safety of the network. The high-pressure water blasts away this wicked hindrance to efficiency. Just like the test train that I'm waiting to see, the RHTT is running along a vast amount of track every day. Unknown heroes of the railway. The RHTT departs with a squeaky two-tone and a wave from the driver. Even though I'm alone on the platform, the interaction gives me company and leaves a smile on my face just from this driver's simple acknowledgement.

A flash of electrical discharge in the distance. I press record on my camera. BMAC light clusters and a central light on the roofline, a distinctive feature of the class 73/9, make themselves clear. They cast across the darkness like a lighthouse as the lead locomotive darts over sets of points, changing its angle of attack, leading it into its designated platform. The colours of the GB Railfreight livery and the trailing yellow UTU carriages catch hues of the surrounding streetlights. I am grinning from ear to ear as it approaches. 'It's 73962!' I knew it'd be leading in but, given the unpredictability of trainspotting, I am still relieved to see it. The loco stays on its path, rocking gently towards the fingers of light reaching out from the station. Then it darts across another set of points. 'For goodness' sake, I'm on the wrong platform!' I got sidetracked when I arrived, looking at 313201 on platform 8 instead of being on platform 5.

While keeping my camera tracked on the test train rolling in, I start fast-walking down the platform, half tutting, half giggling at my own incompetent planning. The test train outpaces me as it rolls in at around 10mph. It decelerates for a final time and comes to a stop as I near the end and circle around the bays of platforms 6 and 7, and up to platform 5, the gateline staff laughing with me. There she is. 'Here we have 73962 *Dick Mabbutt*,' I announce to my camera. 'What a beautiful locomotive she is.' The nameplate calls out to me. It's rough and painted orange in its recesses, from which rise the smooth letters cutting through the orange surface, standing proud of the roughness, showing the blunt silver of their metal. The two contrasting textures and colours make it so satisfying to look at.

The driver jumps out of the cab. I smile at him and ask if it'd be possible for him to take a photo of me next to the locomotive.

'OK, that's fine.' He holds the phone with his arms fully outstretched and snaps a few photos.

'Thank you!' I smile.

I double-check the photos, take a few more of the locomotive itself, then wander up the platform. I peer into some of

the mark 3 coaches where a few of the workers are looking at the monitors with their hands on their heads. The windows are tinted and it looks quite cosy, with a few cups of tea and biscuits here and there. I'm still buzzing from the moment it came in and the unpredictable fast walk that ensued. At the end of the consist stands 73965 *Des O'Brien*, *Dick Mabbutt*'s younger sister. The driver has already hopped in the cab and has switched the orientation of the lights. It's due to stand at the platform for another ten minutes, so I take a few photos of it under the two bosom-like extrusions of Brighton Station's canopy. Standing towards the end of the platforms offers no protection from the wintry winds and I am regretting not wearing a scarf, but the moment is worth it.

A class 377 unloads its passengers behind me on platform 6 just as the test train announces its departure from Brighton with a hushed two-tone. I give the driver a big smile and a wave. Passengers returning home are puzzled by my happiness over this seemingly normal train, to them

at least. *Des O'Brien* gets on with her job and pootles out of the station on notch 3. The coaches sing while the wheels click in semi-quavers: *click-click, click-click ... click-click, click-click. Dick Mabbutt* whirs past and bids me farewell, just like she did at Gatwick Airport. Eyes red, fading off into the darkness.

I set off too, and park under some trees just outside my place of work. The car park is locked. It's 01:47 and very cold. I am happy though, satisfied and content. I recline my seat as far as it can go, pull my knees close to my chest, and close my eyes.

Chapter Four →

The Crossover

Location: Kingston, 23 November 2021, 08:06

Waking up this morning is a bit of a struggle. My eyes feel puffy and are stubbornly refusing to adjust to the light. I force myself out of bed and put on one of my favourite jumpers – a Missoni one that I've inherited from my grandad. It's predominantly burgundy and brown but has all the colours of the spectrum running through it in a grid format. My grandad has lots of colourful jumpers and once sent me a package with some of the ones he didn't want any more, including this one I'm putting on now. I pull on my extremely baggy Levi 569 jeans, my salmon pink Clarks Wallabees and my brown puffa jacket before setting off outside.

Lucy sits waiting on the drive. I hop in, quite happy to find some orange and lime tic tacs I left there on my last journey. It's a pleasant morning, not overly cold, quite warm in fact for November.

Today I'm heading to Gloucester Station. It's a busy day there today: a class 37 scrap move, the beautiful steam-powered *Duchess of Sutherland* on a railtour, and a class 43 NMT, all passing through twice. The car journey takes around two and a quarter hours. On the way I tune into Radio 4, my go-to radio station for a long journey. Along the M4, there's an interesting discussion on CRISPR-Cas experiments, with a professor from the University of Nottingham going into a lot of depth about genome editing. Much of it goes over my head but it helps my mind to concentrate on driving while also allowing my brain to idle a little on snippets of the conversation.

Location: Gloucester Station, 23 November 2021, 10:47

It was a clear morning in London, and I was hoping it would be a similar case in Gloucester, but as I pass Swindon I enter patches of fog. Across Gloucestershire there is cloud cover, albeit light and wispy. The station car park is quite busy this morning. There's one spot that's free along the wall furthest from the tracks, in a row of spaces occupied by mounds of amber maple leaves about 10 centimetres deep. It seems that the sweeper for the car park has done an excellent job of keeping it leaf free but has subsequently created a substantial pile that runs across multiple spaces, meeting the sills of the cars that have parked on top. Opening the door and placing my foot on the ground, there is a satisfying amount of crunchy cushioning before it reaches the floor. Traipsing through the leaves and playfully kicking some up in the air, I pay for the parking, day rate only, which for once is quite useful, as today I will keep my car here for the whole day rather than only for a few minutes.

The gates to the platform are open. I walk through and just to my left see Ryan with his old friend Darren. Ryan asked me beforehand if I minded whether Darren joined us. 'Sure!' I replied. I still feel worried though. I have never met Darren before so I don't know what his presence will bring to the day. Sometimes, if someone's energy feels a bit incongruous – usually if they're too loud – it can throw me off and I can retract into a bit of a shell. What Ryan and I have when we are together is dynamically special. Now that we know each other well, Ryan is at ease with my presence and I'm at ease with his. We both get excited about the same things and can also be calm and quiet too, enjoying each other's company without having to fill every silence.

I approach the two of them. Darren is wearing very bright white Air Max 95s, skinny jeans, a white Adidas top and a big puffy jacket with a faux fur hood, I assume he is around the same age as Ryan, 31, as he said they used to go clubbing

Class 373

Build date 1992–1996

Total produced 38

Number in service/preserved 11, remainder scrapped

Prime mover Traction motors: Brush TM2151B

Power output 25kV: 16410 hp, 3000V: 7600 hp, 700V: 4600hp

Maximum speed 186mph

Current operators Eurostar

together. Ryan is wearing desert boots, waterproof trousers, his trademark bright blue waterproof jacket with his CrossCountry badges on, CrossCountry lanyard and a red backpack. There's also a yellow Morrisons bag on the bench next to him, which I assume is his. As I get nearer, Darren pulls a snood over his face. I think to myself that it seems respectful that he's considered covering his mouth and nose during these Covid times. Ryan is fiddling with his tripod.

I smile to them both. 'All right, guys!'

'Alreeet,' Ryan replies, looking up then looking back down at his tripod.

Darren puts his thumb up, and I can see from his eyes that he's smiling. 'Yo, I'm Darren, nice to meet you, man,' he says.

'Nice to meet you too,' I reply, still smiling. My first impression is that he seems kind and genuine. He's not a train enthusiast but I think he just wants to hang out. I relax a little.

We start to make our way down the platform to prepare for the arrival of the LMS Princess Coronation Class 46233 *Duchess of Sutherland*, a steam locomotive built in 1938, 4-6-2 configuration, with muscular shoulders and now, after a repaint in 2018 to celebrate its eightieth birthday, in a dashing crimson livery. I catch up with Ryan and Darren after lagging behind. I think they're talking about my train-spotting videos as I jump in midway into their conversation.

'The lockdown last year, what else was people doing?' Darren asks rhetorically.

'You inspire me to keep at it, Francis,' Ryan chimes in, looking at me. 'I'm not gonna give up until I—'

Darren interrupts: 'Yeah, you never liked trains much, did you?'

Ryan turns and looks at Darren with a screwed-up face. 'Yeah, I did, I've always loved travelling on trains, Darren.'

'Well, your inspirement has gone up since Francis's videos, hasn't it?'

'Yeah, well, he's inspired me to not give a damn about what people think.'

I smile and turn to Ryan. 'Yeah, you've helped me to keep at it too.'

Ryan looks at Darren and nods with a smirk.

We establish our position on the platform about 30 metres from the end, in the direction of Horton Road crossing. Ryan has brought his old phone and camera to set up on a pair of tripods, which pen us in from either side, both facing outwards, leaving us 5 metres in between them. If we cross the line of sight of the camera then it'll potentially ruin Ryan's shot. This is something both Ryan and I strive to stay aware of in more crowded situations with lots of cameras and I'm totally aware of it now for the sake of Ryan's footage.

From further up the platform I see someone strolling towards us. I squint. 'Who's that?'

Ryan looks up. 'Oh, that's Jake. You all right with him being here?'

'Yeah, sure.' Jake and I have met before, he's in his early twenties and more of a fan of steam engines – or 'kettles' as we call them jokily – than diesels so I should've expected him to be here today.

'How's it going, all right?' Jake announces.

'All right, Jake?' I reply.

'When's she due in? Soon, I bet.'

'Ten mins or so.'

'How's the missus doing, all right?'

'Yes, she's very well, thanks.'
'Gooood.'

The *Duchess of Sutherland* has to power up the slight incline to Horton Road crossing – *puff puff puff puff*; we can see the steam floating up over to the top of the tree line. Adrenalin starts to pump around my body. I can tell Ryan is feeling good about it too. It sounds so soft from a distance, the releases of pressure just seeming to stroke us as the power of the locomotive is muffled by the surrounding industrial clutter and road noise. Its gloss-black face proudly leads around the corner – so bold. It finds its path into the station after passing over the points and a level crossing. The power cuts off. It starts to coast. *Click click click click* – the puffing ceases but the rods and pistons still clunk along below the smooth metal panels of its boiler.

'That is beautiful,' Ryan whispers.

We stand stock still as though a predator is prowling by. Will it notice us and let off a whistle? The tension builds as the train approaches. *Click click click click*. The front of the LMS Coronation Class coasts by, no whistle. I turn to Ryan, who's shaking his head in disappointment. Ryan loves whistles, as do I, but it often seems to be a make-or-break situation for him when a locomotive does or doesn't give a whistle. I smile at him to try and gauge how he's feeling. He's busy looking at the loco, still passing us.

The bodywork of the *Duchess* strikes me: the cylindrical boiler grows muscular just before the footplate and seems to dip downwards, almost a reflection of the direction of the energy transfer, from the firebox, along the boiler and down

Cepheus

60019

DUCHESS OF SUTHERLAN

365509

to the piston. The footplate goes by and I briefly make eye contact with the fireman. I don't think I know him but he looks friendly. The crimson livery of the *Duchess* matches perfectly with the rolling stock owned by West Coast Railways. Inside, each table has a lamp with a dangly shade, illuminating white tablecloths set with silver cutlery. The demographic of those on board is generally sixty-five-plus, many perhaps remembering the days of steam power from when they were children, before diesel engines became the dominant source of motive power in the 1950s. Mark 1 carriages with compartments, more connected to fellow travellers, a more visceral experience; windows that you could open and shut by unclipping a mechanical pin and sliding the glass; cloth on the headrests of seats to prevent gentlemen's hair product from staining the seat fabric; the sound of a mechanical animal hauling coach after coach through the harnessing of boiling water. It's a completely different world for us train enthusiasts now.

I recommend to Ryan that we make our way down to the other end of the extremely long platform. He shakes his head and points out his cameras: the coaches are still rolling by, he still wants to complete the shot. I whisper to him, 'We need to get down there, it's not here for long.' I don't want to speak over his video and I know he's taken particular care with this one. Ryan seems frustrated, looking left and right. Then Jake legs it down the platform, right in front of Ryan's shot, as the third from last coach passes. Ryan accepts his fate and picks up his camcorder.

'I'm going to head down, Ryan. Are you coming?'

'Yep, bear with me. Darren?' Darren, leaning against the fence, suddenly leaps into action. 'Darren, can you help me pack up my other tripod, please?'

'Yep!' Darren has picked up on Ryan's sense of urgency.

I start fast-walking down the platform. The train is still cruising and hasn't applied the brakes yet. I look to my right and see some elderly couples laughing at us scrambling to keep up but limiting ourselves to a fast walk while lugging our kit. Jake is laughing his head off as I catch up with him, cackling, making me laugh too. I look right again and the coaches have slowed to the speed we are walking at. Through the window, a man makes a running gesture with his arms while his partner laughs. I laugh harder while navigating around other passengers near the station building. Ryan and Darren have caught up. Following around the curve I can see the hive of fellow train enthusiasts surrounding the loco-motive. There are around twenty-five of them trying to get a good shot of her, a lot of old heads with white or no hair and young heads with iPads and smartphones. I navigate around

the edge of the crowd and try to sneak a shot of the *Duchess* at the edge of the quarter circle. Steam rises from between the bodywork and the edge of the platform, giving perfect ambience to the footage I am filming. I switch places with Ryan as I know he's keen to get a video too. Retiring to the edge of the crowd, I peer into the footplate and the fireman gives me a wave. I wave back and smile.

Green! Suddenly the signal changes its orientation and I quickly dash around to my original position near the front of the engine, its cylinder on the left side bulging right next to me. *Phhhhwumph.* The first puff of steam leaps from the chimney. *Phhhhwumph* – another energetic puff. *Wwww-wOOOOOAHHHHH WUP-WUP!* A tremendous tone from the whistle grows until it's at maximum volume where it's highlighted by two succeeding chirps. I cry out in reply. The locomotive continues to puff as the exhilaration pumps up and down my arms and courses through the rest of my body. Departing with a wave, the fireman on the footplate nods too with a big smile across his face, well aware of the joy he is delivering. The passengers wave too, hands lifted, excited gestures from children and a general exchange of warm energy all facilitated by a metal firebox and boiler. This is why I feel trains are so much more than just trains: they are able to bring people together, make people smile and create lifelong enthusiasm for special moments like this one. Modern trains are slowly drifting away from the essence of what makes the railway so enticing, the drive for efficiency slowly dulling the human-like characteristics and imperfections that give vintage locomotives a character and a soul.

Smiles abound as we saunter back to our end of the platform and stand by for what is, for me at least, the main event. This is what I came for: the class 37 scrap moves. Recently,

Signals

On the UK's railways, two types of signalling are used. The less common signals are semaphore signals that use a mechanically operated red arm with a white stripe to signify to the driver whether to proceed or not. Attached to the arm is a red lens for one orientation and a green one for the other, for use at night. If the arm is in the horizontal position with the red lens illuminated, the driver should not proceed. If the arm is lowered and the green lens is showing, the driver can proceed. These are called semaphore stop signals. Semaphore distant signals are used, too, with a similar function, but instead of the red arm and red lens for 'stop', it is yellow, with a notched arrow at the end. Instead of instructing the driver to stop, while it's in the horizontal position the arm and yellow lens (for the semaphore distant signals) instruct the driver to 'proceed with caution', meaning the next signal could be red.

The more common signalling system in the UK comes in the form of colour light signals. Each light has a different meaning:

Green	proceed at line speed or if moving at less than line speed, the maximum speed of the train
Double yellow	the next signal is displaying a single yellow
Flashing double yellow	the next signal is displaying a flashing single yellow
Single yellow	the driver must be prepared to stop for the next signal
Flashing single yellow	the driver must slow down for a diverging route that is lower than line speed
Red	Stop!

Rail Operations Group have been utilizing their fleet of class 37s to drag retired class 365 units from their long-term storage site at Long Marston in Warwickshire to the scrap-yard at Newport Sims, where they are stripped down to scrap metal. I have mixed feelings about this. On the one hand, the class 365s are trusty units. They served our network for twenty-five years and recently underwent a makeover that gave them a new face, which really does look like they are smiling at you as they approach. However, pre-facelift they did look better. (I am very fond of the old 'networker look'. I even had it as a 00 gauge model in the old Network South-East livery.) They are great units and they will be missed; I'm sad to see them go. On the other hand, the scrap move feels like a historic moment and I cannot wait to see today's class 37 performing this one. It's 37884 *Cepheus*, driven by David, who Ryan knows through Facebook. He's a great driver and a railway enthusiast too. I imagine he's more likely to be a thrash merchant, or in other words someone who thrashes the locomotive (sensibly). He is just passing through Chel-tenham Spa and is coming in hot. It has been a long time since I last saw a class 37, so this should be great.

The rumble of the type 3 English Electric traction motors electrify the air as it passes Barnwood Junction. 'Oh my God, it's thrashing!' I call out. Ryan is beaming next to me. He has only one tripod set up this time as he found having two impractical, and the one he has set up is facing down the platform so I don't have to worry about interrupting his video. 37884 emerges from around the corner, its lights akin to a beaming smile. I'm nervous about what's going to happen – it's pure anticipation and excitement. I'd liken this moment to the beginning of a fireworks show on Bonfire Night, when you know there's going to be a display but don't know what the fireworks are going to look like, the order in which they will happen or when they will explode. The class 37 is moving at a snail's pace relative to the frequency of my heartbeat and the feeling in my body. It feels like time is slowing down.

Ryan is narrating for his video: 'This should be David driv— OH MY GOD, THERE'S TWO!'

'AWHOHOH!' I jump in the air and clasp my hand over my mouth. A class 60 pulling an empty Murco oil train service is approaching platform 1 while the class 37 is going through the central avoider line. To put it simply, they're drag racing right next to each other on different tracks. It's hard to highlight how special this is … it's incredibly rare. 'No … no … no way, there's the Murco lining up next to it!' The class 37 has already crossed the converging points on to the UML, but the class 60 on the Murco is only just lining up on the track into platform 1. By the time they're both on the straight path into the station, the Murco is just at the back of the last class 365 unit.

'The Murco's drag racing!' Ryan shouts.

37884 approaches the beginning of the platform and lets off a powerful series of tones: *DUU DIIII DUU … DI DU*. I jump up and down.

Ryan shouts, 'Thank you, David!'

'Thrash, please?' I call out, hoping to hear some from the locomotive as it's been a very long time since I last had all-over-the-body goosebumps from a type 3 English Electric.

David lets off an Ilkley Baht 'at tone – *DU DI DI-DI DU DI* – so-called because it's reminiscent of the intro to the Yorkshire folk song 'Ilkley Moor Baht 'at'.

'Thank you!' I shout. 'Oh, and the Murco is thrashing!'

The class 60 opens up the taps. A cloud of heavy black clag

rolls out from the top of the locomotive; with the tanks it's carrying not being full of fuel, it's able to tug the reduced load at a greater acceleration. The class 37 is coasting through the central line, but clearly the class 60 is going a lot faster, eating up the rear carriages of the class 365 unit, which the class 37 is dragging, as it passes it on the right. David, at the helm of 37884, waves as it rolls by. 'Hello, David. Oh, look at it – *Cepheus*.' My mind can't comprehend what has just happened and I can't get my words out. The class 37 passes. 'And the Murco! Woah!' It has picked up some serious speed now. A class 60 thrashing isn't as impressive as a class 37, it sounds like a large industrial fan with a low-level rumble underneath. Still, it wasn't the sound that made this particular moment so satisfying to watch, it was the sight of the two trains on directly adjacent tracks, going in the same direction, at different speeds. Usually, on networks around the

UK, tracks next to one another will run in opposing directions, even when there are stretches with five or six tracks abreast. Luck was on our side though: if the class 60 had been ahead we wouldn't have been able to see the class 37 on the scrap move, running on the tracks behind it. When this happens, it's called getting 'bowled'.

60019 passes in front of us, a brief wave from the driver as he whirs by, the clag from the exhaust slightly less sooty now. 'He's bowled us!' I laugh, half joking. This is not quite complete bowlage but there's no chance of seeing the rear of the smiley class 365 now, the unit being dragged to the scrapyard by David in his class 37. I didn't even get to see its number! Instead, the Murco's grimy oil tanks fill our view. Most are long and brown, once presumably silvery grey, some ribbed, others more sleek. Newer tanks are painted red and green but are still filthy. Patches cover ancient graffiti. Most of them are stained and weathered particularly where the fuel has been loaded. In between the speeding tanks I catch glimpses of the class 365's comparatively cleaner white Great Northern livery. It definitely seems to be slowing down.

'Get out of the way, quickly, I want to say goodbye to the 365s,' I say jokily to Ryan. His face is creased with amusement. The white of the Great Northern livery peeps through again. 'Ryan, it's still here, look, it's still here – run along, run along.' I can see his facial expression change as his smile is replaced by a look of concentration, as though he's thinking about what he should do next. We might get another chance at seeing the 37 so I start making my way down the platform. I turn around. 'Ryan!' He looks at me with a 'YES!' while fiddling with his tripod. It's such a high-pressure situation. A good video depends on it and Ryan's tripod, with all three 1.5-metre legs extended, will take at least a minute

Freight Movements

The majority of trains on the UK's network are passenger trains. They are powered by an exciting, varied group of engines, from class 43 HSTs, class 91s on the East Coast Mainline and class 68s on the Chiltern Mainline, to the class 67s on the WAG (Welsh Assembly Government) trains in Wales. A common feature unites these trains: they're all locomotive-hauled services. This means that the carriages are not individually powered and rely on the locomotive(s) at the front or front and back to move them from A to B. Much like a conventional steam train. This requires the locomotives to have monstrous engines, producing thousands of horsepower and thousands of newton metres of torque. Big rumbling units, shackled to the body of the locomotive.

They sound fantastic, particularly the older ones with little silencing on the exhaust. For TOCs (train operating companies), typically locomotive-hauled services are less fuel efficient and will suffer from more traction issues in comparison to their multiple-unit counterparts, where carriages contain individual power units that, in unison, move the train. This means locomotive-hauled services are becoming less common.

Fear not. Freight trains, treatment trains and test trains are all locomotive-hauled, and in most cases they have thousands of tons more to carry, which means more thrashing! Most enthusiasts will agree, freight is far more exciting than passenger services. There is plenty to see if you know where to look. A 3000-ton aggregate train might blast through the station whilst commuters are waiting for their train home. Aviation fuel carried in huge tanks can be seen carefully navigating points. And there are the railhead treatment trains, blasting away leafy residue from the railheads, preserving adhesion on the rails for passenger trains and keeping services safe. An almost endless amount of these movements occur every day under our noses, keeping our economy moving. Even more fantastic is the fact that they are hauled by the workhorses of our railway: the class 66s, class 70s, class 60s, class 56s, class 37s, to name but a few.

I have compiled a list of some of my favourite non-passenger movements, where to see them and a bit of information about them.

For timings check live.rail-record.co.uk

Service	Locomotive(s)	Purpose	Weight	Spotting locations (my personal preference)
Whatley Quarry to Dagenham Dock	Class 66, Class 59	Aggregate transportation	4,400 tons	Westbury, Bedwyn, Reading, West Hampstead
Bridgewater to Crewe	Class 68	Nuclear waste transportation	715 tons	Bridgewater, Bristol Temple Meads, Stafford
London Paddington to Derby	Class 43	New Measurement Train (NMT)[1]	N/A	London Paddington, Taplow, Severn Tunnel Junction, Tamworth
Baglan Bay to Chirk	Class 56	Timber transportation	1,200 tons	Baglan Bay, Newport, Hereford, Shrewsbury
Willesden DCR sidings to Machen Quarry	Class 60	Empty wagons to fill with aggregate	600 tons	Southall, Reading, Bedwyn, Severn Tunnel Junction
Stewarts Lane to Newport Sims	Class 66	Dragging Southern class 455s to the scrapyard[2]	600 tons	Kensington Olympia, Reading, Severn Tunnel Junction, Newport
Bristol Kingsland Road to Derby	Class 37	Ultrasonic Test Unit (UTU)[3]	N/A	Bristol Temple Meads, Bromsgrove, Birmingham New Street

1. The New Measurement Train (NMT) is comprised of two bright yellow class 43s and five equally yellow mark 3 carriages. It uses a laser track scanner, track geometry system, high-resolution video camera and an unattended geometry measurement system. These systems prevent faults such as: Twist – when the rails don't sit in line with one another, Cyclic top – where dips in the rail surface can cause the train wheels to bounce, which over time increases the severity of the dip, Gauge variations – where the rails can deviate from the 1,435mm width of our network. Other systems used on the NMT: overhead line inspection systems, PLPR (plain line pattern recognition system, which compares what it's seeing to what the track should ideally look like) and a radio survey system which tests that the GSM-R (Global System for Mobile Communications-Railway) in the area is working as it should be.

2. The Southern class 455s destined for the scrapyard cannot drive there under their own power, as there isn't third-rail power after the West London Line, therefore they need to be dragged by a diesel locomotive. Grim Reaper runs, as they're called, are certainly bittersweet: a last chance to see a multiple unit but also an interesting move to watch. Newport Sims is a good location to check on live.rail-record, as there are plenty of other locations that will be turning their stock into baked bean cans soon.

3. The UTU uses ultrasound, so tracks can be examined for any cracking internally and externally. It also uses radar that penetrates the ballast to look for voids up to two metres deep, and a laser rail profile scanner that compares the railhead to what the profile is expected to be. If any issues are found, a tamper track machine can come and readjust the balance of ballast, or if the railheads are worn undesirably, a rail grinder can restore them.

to retract. He points down the platform to Darren. I can't hear what he's saying but he's most likely instructing him to get moving. Ryan bunches his tripod legs together, lifts it off the floor and accelerates towards me. I turn and follow the direction of the tankers at speed, Jake right beside me. We must have been passed by at least fifteen tankers and I turn to see that the last one is about to overtake us. Focusing on Ryan, I have taken my eye off the snatched sightings of the class 365s so I have no idea where they are. The last tanker is red, barely, turning dusty brown around the edges of the tank and bogies: the cast steel frames that hold the wheels in place and connects them to the body of the tanker. The rear light flashes red at us as it passes – and there it is! The

Class 60

Build date 1989–1993

Total produced 100

Number in service/preserved 29 in service, 66 stored, 3 preserved, 1 scrapped

Prime mover Mirrlees Blackstone 8MB275T

Power output 3,100bhp

Maximum speed 60mph

Current operators British Rail, DB Cargo UK, DCRail, GB Railfreight

Nicknames Tug

smiley face of 365509 is revealed, crawling along on the set of tracks behind.

'HE'S STOPPING!' I shout.

We all accelerate. Jake starts growling like a dog while moving as fast as he can. David, the driver, is on a yellow signal, which must be why he's going slow. We momentarily catch up with the dragged unit, drawing about level with the rear of the train as we match its pace for around ten seconds. But then we start to slow. 'Don't go!' Fifty metres of platform is tiring us out; we're no match for the class 37 that isn't even trying. I think it might be accelerating too. It starts to pull away, and Jake blurts out in a rattly voice, 'Oh, I ain't gonna make it.'

I turn to him. 'Come on, we can!'

I see that everyone has been reduced to a walking pace. My feet slap against the floor as I slow down in a bundle of inhalations and exhalations. The set of 365s moves forward, opening up the curvature of the platform to my line of sight. 'Hang on!'

Suddenly the realization hits. The centre line, the line the 365s are on, converges with the line running past platform 1, carrying the Murco tankers, which are just pulling clear of the station. The 37 will have to stop to let the Murco through and leave enough time for it to get two signal blocks in front. 'He's slowing down!' I shout to Ryan.

Ryan peers around the corner. 'He's got a red, he's got a red, he's got a red!' He accelerates — we all have a sudden rush of energy. I focus on my breathing and my running like I'm partaking in a mid-distance race, conserving my energy but also trying to get there before it departs. Three hundred metres left of the platform — come on! 'Yes, yes!' I mutter through my heavy breathing. The class 37 and 365s have completely stopped at the signal, I just need to get up to the front. Echoes of the idling type 3 English Electric power unit travel under the canopy towards me, pulling me closer.

The curve to the right conceals the class 37 until I've passed the second carriage to the front; its yellow face then pokes out, still stationary. Ryan is behind me, now only holding one leg of his tripod: the other two have splayed out, by the look of things making it quite a hard task to carry without bashing into people on the platform. Now that I'm right next to the locomotive I notice for the first time that on the nameplate 'Cepheus', the name of the locomotive, is accompanied by the star constellation after which it is named. Individual stars raised up from the flat metal, like silver pin badges. The letters of the word 'Cepheus' are raised from the red background of the nameplate in a boldly contrasted silver. It fits with the livery so well, dark grey with a phoenix rising out of red flames, Rail Operations Group smartly dressed in the centre.

The door opens, and David's second man looks out and sticks his thumbs up with a smile. A hand reaches up to the top of the driver's-side window to press down the latch and pull down the glass, the tops of the fingers emerging to draw it to halfway, then the palm to push it all the way down to the bottom before the hand converts into a thumbs up and David comes into view with a smile on his face. He's a bearded man with glasses, looks quite tall too, in his late forties or early fifties I'd say.

'Hi there!' I shout over to him.

'All right?' I can just about hear his response from the cab but his attention is on what's ahead of him.

Ryan catches up with me, breathing heavily over my shoulder.

Before we can regroup the driver's door is slammed shut and pressure is released from the brakes, *tshhhh*. 'Oh, they're off,' I say, surprised. David rests his forearm on the side of the window like an American film star driving a convertible Ford Mustang.

'I am out of breath, don't pace it down the longest platform,' Ryan says, just as the locomotive pulls away.

The first round of thrash pulls the 150-ton unit into motion. It fires hot clag up into the air as combustion from the locomotive fights with the cold calm of the station. 'Ahhhh' is my involuntary reaction to the portal of thunder David has just opened up. Thrashing stops momentarily to prevent wheel slip then the second round is induced with a brief two-tone and another thumbs up from David. *DU DII*. The rpm builds up, growling, the turbochargers spool, whistling

intensely but at a delay to the onset of acceleration. Exhaust gases blow through the turbochargers, driving a turbine at tens of thousands of revolutions per minute that then drives a compressor that sucks, compresses and discharges air to an intercooler and then into the combustion chamber at a higher pressure, feeding air into the engine and increasing the power output. This dependency on exhaust gases to drive the turbo causes a delay in the time between increasing the rpm of the engine and an rpm increase in the turbocharger. That's why the screaming of the turbocharger builds up slightly later. It is an epic sound. On every exhalation I can't help but make a noise as my body connects with the thrashing of the locomotive, the vibrating air causing the hairs on my arms to stand on end.

'Oh yeah, HELLFIRE!'

The class 37 rips up the very slight incline out from Gloucester Station. It powers down as the elevation plateaus

---

Britain's Oldest Station

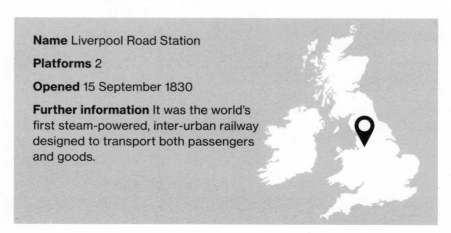

Name Liverpool Road Station

Platforms 2

Opened 15 September 1830

Further information It was the world's first steam-powered, inter-urban railway designed to transport both passengers and goods.

over the bridge. 365509 smiles at me for the last time as it follows the Grim Reaper to the mechanical claws, the first step of its recycling process. The track leads downhill towards Over Junction. Gloucester Cathedral stands tall next to the disappearing train, the last frame of this beautiful encounter.

Chapter Five ➜

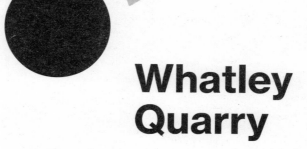

Whatley
Quarry

Location: Frome, June 2006

Before properly starting at school in Somerset, I had a taster day. Wearing an orange and white checked shirt, I ran around the playground in my own world, making explosion noises with my mouth. Everyone else was in a red school uniform, I was not. Some of the older pupils were whispering and giggling about me but I didn't care. I reasoned that maybe the giggling was because one of the girls fancied me. I smiled and continued blasting around the playground, skipping and jumping. It was one of the last days of school before the summer holidays, a swelteringly hot day, the teacher even asking one of my soon-to-be classmates to get me a cup of water because we needed to stay hydrated. He hopped on one leg to the tap and returned with an orange plastic cup; approaching me, he spilled some of the water on to my top. 'James, watch what you're doing!' said the teacher. I smiled at James and he sat back down. The air felt cleaner than in London. All the classroom windows were open, letting in the breeze and the sound of the swaying trees. I remember feeling happy, even though I didn't know anyone. I would go home in a few hours to run around in the garden with my brother and play with my trains. I was full of optimism. In September I would be joining a Year 2 class.

Primary school only lasted three years for me, joining in Year 2 and leaving at the end of Year 4. There's a funny middle school system in the area that bridges the move from primary to secondary education, from Year 5 to Year 8, so there wasn't much chance to get settled in any one place. Still, my time at primary school was great fun. We had a teacher, Mr Balmer, who had a very relaxed teaching style and always made our lessons interesting and enjoyable. He was often assisted by his puppet, Mr Frog, who would pop out from the cupboard and deliver a two-minute Kermit-esque monologue on a range of subjects. Once, I brought Mr Frog home, thrilled to show him to my brother. On another occasion I was crowned the winner of a competition to spell the longest word, mine being 'encyclopaedia', for which I was rewarded by being allowed to go to break time two minutes early.

James became my best friend. He used a wheelchair, and during break times we would pretend to be co-pilots in a fighter jet: I'd push him around at considerable speed and we'd talk over our pilot headsets, coming into land and accelerating for take-off over the bumpy concrete near the roots of the trees. We were granted special access to the sensory garden, where there was a long concrete path leading up to a sensory circle that looped around the edge of a deep flower bed. Presumably the teachers thought we would be having a pleasant time admiring the flowers, but actually this was the perfect runway and flying loop for a Concorde to complete a trip from London Heathrow to JFK International. We would line up at the end of the runway. 'Three, two, one, take-off!' I would make Rolls-Royce Olympus sounds over James's shoulder. 'Airspeed building ... V1 ... rotate ... positive climb ... gear up.' By the end of the straight, entering our loop, we would nearly be at Mach 1. 'Ladies and gentlemen, I hope you are enjoying your flight. If you look to your right you will see we are just passing the English coastline.' The difficulty wasn't holding the speed at this point, but keeping us both in a constant curve. My wrists would feel the strain, pushing one grip and pulling on the other while maintaining a forward motion. Once, inevitably, James's innermost wheel left the concrete and dropped into the flower bed, where the increased drag brought his wheelchair deeper into the border, unfortunately tipping it over. I gasped as I watched him fall out. He laughed off the incident and was absolutely fine, but that marked the end of our flying adventures. From that point on I was banned from pushing James's wheelchair.

Through primary, middle and secondary school, my model railway grew, never finished: there were always new areas to develop, longer stations, more interesting trackwork. My hands became more dexterous so I was able to experiment with more detail. I became more engrossed in my world. My brother got involved too. We were given a camcorder for Christmas or a birthday, I can't quite remember which, and we used to make films with it: comedy sketches where we'd hit each other with pillows; POV videos where we would put the camcorder in our Action Man car and launch it off Action Man-sized cliffs. On the model railway there would be a runaway train or a car that had become stuck on a level crossing, only for a class 142 to smash into it and the people trying to help, all the while with our background commentary and sound effects, usually screaming and shouting. We would watch these videos back on the minuscule LCD

screen, the tinny audio hilariously distorting our shouting. We would laugh our heads off, show our parents, then try to make something funnier. It was an amazing time, the sand box in which my brother and I could grow our creative muscles.

Frome Model Centre was the fuel for my model railway fire. It sits on the top of Catherine Hill, a very steep, narrow cobbled street. I once smashed my forehead into one of the cobbles when it was icy and have approached with caution ever since. The cobbles meet normal tarmac at the top of the hill and the road opens out, while the sturdy old pavement draws around the edge of the model shop. There's a canopy that hangs over the entrance, looking very similar to the sort you can find near the edge of old platforms, featuring a repeat-pattern valance running around the perimeter. Accents of green surrounded the cream paint that always seemed to be in the best nick with no flaking. Walking in, a bell jingled as the single-glazed door was pushed and there was usually a nice lady who said hello. Just to the left was where I used to find what I was looking for: different types of model grass, ballast, miniature people, station buildings. Sometimes my mum would buy a small present for me, other times I would just go in to look. Going home with something from the model shop was very exciting as it gave me a reason for further experimentation. I could immerse myself in my world for a few hours, integrating the new addition. This helped to supplement the lack of close rail traffic in the area. I hardly ever saw actual trains.

We had lived in Somerset for a few years before I realized there was heavy freight traffic nearby. Whatley Quarry, the biggest carboniferous limestone quarry in Europe, is a few miles outside Frome. Most weekdays there's a siren from the depths of the quarry that emanates into the soundscape surrounding the town, followed by a ground-rumbling blast that removes thousands of tons of rock. These chunks are picked up by huge loaders that release the rubble into behemoth dump trucks that wind their way up to the processing facility. Once broken down there, the aggregate is loaded on to trains that wind their way to distribution hubs around the UK, before it finds its way into various parts of the construction industry.

This was all made known to me when the place where my brother and I would ride our bikes was buried under aggregate. A train loaded to 1,700 tons became runaway after the

handle used to operate the train's brakes failed, the slight decline away from the quarry proving to be enough to overcome the brakes on the shunter too, which was manoeuvring the train at the time. Picking up speed, the shunter's high roofline collided with the tighter mouth of a tunnel just after the siding, compacting to the height of the opening and continuing through it. After just over a mile, the consist collided with a locomotive heading away from the quarry in the same direction. The relative impact speed was approximately 15mph – enough to derail the shunter and the following five wagons, which spilled their contents down the bank, over the riverside path and into the river. Front pages of local newspapers carried images of overturned JNA wagons (the classification for open-top ballast wagons), bogies in the air and aggregate everywhere. It shocked me, especially when Dad told me it was just around the corner. I asked to go and see it but he said it was all closed off. After that incident these newly discovered quarry trains fascinated me, but they predominantly ran during the week and I had no idea when. They became mythological creatures to me. I'd occasionally spot them running through the town. The rest of the time my set-up at home kept my railway enthusiasm palpable. The models appealed to the part of my brain that enjoys the linearity of rails and they also satisfied the sonic side as well, with those miniature sounds *click clack click clack* over the tracks. However, during secondary school, even this passion was challenged.

Moving into secondary school was a difficult transition. James and I moved together into the new school but he wasn't in any of my classes, only my tutor group. During break times and at lunch it was nice to have a close friend with me but everyone else seemed to have large groups of peers from their previous schools that they could huddle together with, while we were outsiders. James, my model railway and my cats were my closest companions. I'd look forward to getting school over and done with and coming back to play with my train set. Making new friends was tough, particularly as no one shared my interest in trains, so I decided it'd be best if I listened and watched what people were interested in and tried to adopt those hobbies myself. Mountain biking, scooting and BMXing were all quickly picked up and I started going to the skatepark after school. Over time, I established a nice group of friends and I was a lot happier. Through observation, I started to understand what was deemed 'cool' by my peers and tried to emulate that so I could be even more secure in the group. Not model railways, but streetwear, gym and rugby. I experimented with these, enjoyed it, and felt more confident. I missed my model railway, but it felt like a part of growing up that I needed to do, leaving my childhood behind. The pieces were eventually bubble-wrapped and put in boxes. I still watched train videos on YouTube, still liked to see class 43 HSTs, still loved travelling on the train, but my passion for them was submerged, below the surface.

In 2016 I regretfully sold my model railway on eBay. My 0-6-0 pannier tank, my Eurostar, all the little people and buildings. Taking pictures of them for the listing was the hardest part. I lined up the locomotives and units next to each other as if they were in the sidings together; it reminded me of the way I'd lined up my Hot Wheels cars on the window sill in Harlesden. I debated keeping my GWR 0-6-0 pannier tank with the broken clip on its shell, my first locomotive, but felt that I should let go. The money from the sale went towards new clothes and extending my gym membership. It felt like the right thing to do.

The flame never died out, though. Driving around Somerset, passing under train bridges; passing over the quarry line on the way to school, occasionally seeing a quarry train, a class 66, sometimes a class 59; driving to Bruton and passing through Wanstrow, wondering what the seemingly lesser-used line could be – I was always curious to know where these lines went. Following them on Google Earth, I slowly built a picture in my mind of what the surrounding railway was like. Realizing the line to Merehead Quarry links up to the East Somerset Railway, which would have continued to Shepton Mallet and Glastonbury before being chopped under the Beeching axe. When the BMX pump track was built in the centre of town, I would head down there after school most days of the week and hit the jumps on my bike. Whenever a quarry train went by on the other side of the allotments, I would stop and watch. GB Railfreight's 66718 screeched through once, with its distinct London Transport Museum livery: black underlayer surrounded by diagonal outlines of the modes of transport in London, block-filled with orange, yellow, blue, green, all different colours. It was hauling a load of JNA wagons to Whatley Quarry, presumably empty. That evening I watched videos of 66718 on YouTube. I didn't tell my friends. Taking trips back to London was always special, a particular highlight being a ride on 43002 *Sir Kenneth Grange*, repainted in the original Intercity 125 blue and yellow livery and later 43185 *Great Western*, repainted in the Intercity Swallow livery, to and from London Paddington before the HSTs were retired from the Great Western Mainline. Each one of these moments was a spark that kept the combustion cycle of my train enthusiasm ticking along deep inside, even if the throttle input was being suppressed by my coolness compass.

Going to university was like putting on a turbocharger. In the first few weeks I was surrounded by such an eclectic mix of interests, personalities and energies that my perception of what was deemed cool was all over the place. I had moments of crisis, but what started to break down was my feeling of having to mirror what I saw around me in order to fit in. The social pressure that had been holding down my train enthusiasm was released. Trainspotting came back – it came back in full force. While studying, I'd take breaks and go down to watch the trains run along Attenborough Nature Reserve just outside Nottingham. It was HSTs galore there, even those with powercars fitted with VP185 engines. These were special as they were the younger sisters of the Paxman Valentas, the screamers I had fallen in love with back in the canopy of Paddington Station. They didn't scream as much as the Valentas, they were more muffled, but at speed you could really hear the resemblance, far more exciting than the

Britain's Grandest Station

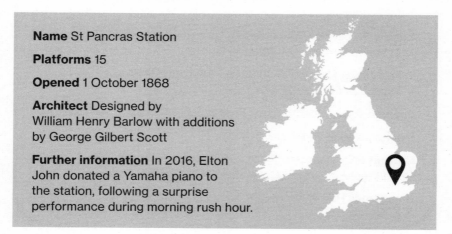

Name St Pancras Station

Platforms 15

Opened 1 October 1868

Architect Designed by William Henry Barlow with additions by George Gilbert Scott

Further information In 2016, Elton John donated a Yamaha piano to the station, following a surprise performance during morning rush hour.

MTU powercars that were a lot more common. The buffered HST powercars were also a favourite, MTU-powered but with interesting modifications from the 1980s. These modifications were needed because of the overdue mark 4 and DVT (driving van trailers) sets that were meant to roll on to the East Coast Mainline with the freshly manufactured class 91 locomotives in 1987. British Rail realized that the mark 4 carriages and DVTs weren't going to be ready in time for the newly electrified line. So, the current mark 3 carriages would continue to be used and eight class 43 powercars would be fitted with buffers for ease of marshalling and remote-control equipment so they could be controlled from the front. In 1991 the eight class 43s had their remote-control equipment

Class 43

Build date 1975–1982

Total produced 197

Number in service/preserved 127 in service, 40 stored, 12 preserved, remainder scrapped

Prime mover Paxman VP185, MTU 16V4000 R41R, Paxman Valenta 12RP200L

Power output 2,250hp

Maximum speed 148mph

Current operators ScotRail, CrossCountry, Network Rail, Great Western Railway, Colas Rail, DATS, RailAdventure

Nicknames HST, Intercity 125

removed and returned to non-surrogate DVT roles. The buffers remained and gave eight class 43s a unique look that earned them admirers. The freight around Nottingham was fantastic too: class 60s hauling the Kingsbury Oil Sidings to Humber Oil Refinery tanker and return; occasional Colas Rail moves – 37s, 56s, 70s going to Nottingham Eastcroft Depot for maintenance. Checking live.rail-record became a part of my daily routine, to see if anything was running in the area that I could go and see. Riding on HSTs from Nottingham to London St Pancras was a regular occurrence too, a chance to switch off and listen to the gentle rumble of the powercar in front of me.

When lockdown came into force in the early spring of 2020, my trainspotting transferred to Somerset. I jumped into the exploration I had suppressed in my teens, picking up the nuggets of interest I had in the area. Cycling to the overbridges surrounding Frome, I'd watch the services from London Paddington blast through on their way to Devon. I searched out where the old Frome–Radstock Line used to connect to the Whatley Quarry Line, finding old relics from the British Rail era. I spent hours researching the old Somerset & Dorset Line, which closed in the 1960s, and how the Frome–Radstock Line once linked up to it, and then traversed the old route to Bath on my bike. The greatest discovery of all was the Whatley Quarry viewing point. I became obsessed with the line leading into the quarry, its route diverging from the passenger line before Frome Station, passing through the centre of town, through Vallis Vale woods, bypassing the old line that it used to take along the river due to the gauging constraints the tighter radius corners posed to today's new, longer locomotives. The track continues along the river once past Great Elm, the section I used to cycle along with my brother. Twelve years later, with the knowledge of the running schedules from the quarry, I recommended times for our lockdown family walks to coincide with when the trains were running, often waiting behind the fence near the tunnel mouth to watch them. I felt a lot more confident introducing my peers to my trainspotting world as well. I would meet my school friend Calum for a bike ride and we would routinely cycle to Longleat Forest via East Woodlands, passing under the Wilts, Somerset and Weymouth railway and the spur to Frome Station. I'd keep an eye out for anything interesting that might be passing through. Calum didn't mind as it gave him an opportunity to catch up and chat about lockdown life while waiting for the trains.

My passion for trainspotting wasn't the only thing that grew in intensity during lockdown: being with my brother reignited the creative bounce that we have when we are together. We made music, a song even about Whatley

British Rail

Founded on 1 January 1948, British Railways, later British Rail, was established after the merging and nationalization of the Big Four British railway companies: Great Western Railway (GWR), London, Midland and Scottish Railway (LMS); London and North Eastern Railway (LNER) and Southern Railway (SR). During the British Rail era, changes swept across the network, including the Beeching cuts, whereby the number of stations on the network was reduced by one-third. Steam locomotives were replaced by electric and diesel locomotives. Passenger numbers surged as journeys were made more convenient by rail. British Rail was then privatized, ending its reign in 1997.

Quarry. We made videos, comedy sketches and even zip lines: I obsessed over the idea of a camera on a zip wire for months, attaching a GoPro to an old scooter body, making the wheels concave and running the wire through the frame. Me, my brother and my dad would trial the filming rig in Vallis Vale woods, breaking one GoPro in the process and thus having to bring into play my engineering studies to try and calculate the desired tensile strength of the wire to support the scooter rig enough so it didn't smash into the floor like last time. I started uploading some of these videos online, and after a while my trainspotting started to become the core of what we filmed. Acknowledging and sharing my passion with the general public was complete catharsis. I was doing the exact opposite to what my teenage self would have done, standing on an overbridge on a sunny evening, watching trains, loving them, filming them and posting the images online without a care in the world. What's more, it started to make other people happy too. My passion for trains had been one of the most secret, most personal aspects of my identity. Now I felt free to express myself. To allow my inner child to be excited and passionate about the railway, to reconnect with the general purity and simplicity of being young and utterly in awe of the sights and sounds. Filming my model trains crashing into cars and laughing with my brother as we shouted into our camcorder has now re-emerged in a new, equally joyous form.

Chapter Six →

Good Tones and Goodbyes

Location: Westbury Station, 27 September 2021, 08:01

The area of Somerset around Pot Lane Overbridge and the Clink Road Junction is my trainspotting heartland. Exploring on my bike during lockdown, I was trying to find the perfect bridge or foot crossing, slowly working my way out from the town. Just outside the A361 near a village called Berkley I discovered Pot Lane Overbridge, and felt like I had found the ideal spot. Standing on the bridge and looking to the south-west, the line flows to the left with swaying fields on either side, bordered by Berkley Lane Overbridge around 300 metres away, just as the track curves and disappears behind the tree line. In the other direction, north-east, the line flows to the right, into a wide expanse of telegraph poles and fields through which occasional foxes bound, sometimes venturing across the tracks. Birds of prey often swoop around overhead. To the left there's a farm, its cows sometimes grazing in the field right next to the tracks, peppering the whole space in blotches of black and white. In the distance, just over 5 miles away, Westbury White Horse shimmers. Trains can be seen coming from that direction only a few minutes

127

after they leave Fairwood Junction, which means they are in view for at least thirty seconds before they pass under Pot Lane. Dark green GWR class 800s, otherwise known as 'cucumbers', frequent the area the most, but there are also GWR class 166/165s and 158s, similarly painted dark green, and the occasional class 166 in the old pink and purple First Great Western livery. Then there are the quarry trains – Freightliner class 66s and 59s and Mendip Rail class 59s. The class 59s and 66s come in an eclectic mix of liveries, my

particular favourite being 59103 *The Village of Mells*, named after the village that the line into Whatley Quarry skirts around on the way in. The locomotive features the bright blue and white livery of Hanson, the cement and construction materials supplier.

@freight_man_59_66_70 is the Instagram handle of my friend Gordon. His posts were recommended to me by Instagram's clever brain, probably because of my incessant searches for #class59 #class66 photos. I noticed he worked out of Whatley Quarry and Merehead, another limestone quarry in the area. I messaged him:

> Hello, I love your posts! I watch the trains coming into Whatley Quarry and I see you are often there! Do you enjoy the route into Whatley, I think it's great?

He replied five hours later:

> Hi, yes I like the route in and out of Whatley and Merehead, I'm normally on the Oxford, Theale or Bassett services, prefer going via the B&H rather than Swindon.

The B&H is the Berks and Hants railway that runs from Reading to Westbury, winding next to the canal through

Pewsey and Bedwyn, which is a particularly favourite section of mine. I replied to Gordon:

> Yeah I can imagine, I find it a lot more picturesque. By any chance are you doing any runs in/out of Whatley/ Merehead on Friday 2nd July?

He came straight back:

> I'm on the Oxford to Whatley, leaves Oxford at 11:45, have to let me know where you will be and give me a wave so I know it's you.

Bridging the gap between enthusiast and driver was a weird moment. Two perspectives on the same industry, one peering through the trackside fence, the other from within the driver's cab. I saw Gordon at Westbury Station, then, after he was held at a signal just after the platform, jumped into my car and saw him again at Pot Lane Overbridge. He gave me some fantastic blasts of the horn, sending my endorphins through the roof, and I was able to make a really great video. After that encounter Gordon became a legend. The few tones that he did and my reaction to them created an energy that was amplified and distributed around the world through the power of social media, making people smile. His simple gesture to make me happy sent ripples emanating out from Pot

Lane, through Somerset and beyond. I think of this a lot. Gordon may not have known that his actions would make so many people happy, but he did it anyway. His small act of kindness for an audience of one spread joy to so many others.

Ryan is on his way to Westbury Station. He's been desperate to see Gordon in action and today I've arranged with both of them for a crossing of paths. Gordon is helming the 08:32 service from Whatley Quarry to Oxford Banbury Road. Ryan's train is due in at 08:10, so I should have plenty of time to collect him and drive us to Pot Lane, factoring in the thirty-five minutes it takes for Gordon to get there from the quarry.

Pulling into the pick-up and drop-off bays at Westbury Station, I come to a stop, put the car in park, push the footbrake in and switch the engine off. Two Instagram messages from Gordon:

> 66524, only a 2 tone horn.

> The quarry wants me out early so keep an eye out.

Oh bugger! He sent the message at 07:23 – he could've left by now! I hastily open up Traksy. *Come on, come on, load!* There he is, two signals out from Frome North Junction. No,

131

this doesn't give us enough time. 'Damn it! Where's Ryan? He must be due in soon!' I mutter to myself. I send him a message telling him what has happened and that he needs to leg it as soon as he's off the train.

> **Ahhh there's no way we can make it to Pot Lane.**

he replies.

> **Maybe the foot crossing before the approach to the station ...**

I suddenly remember there's an overbridge right over the top of Fairwood Junction, just before Westbury. That could work ... I do some investigating on Google Street View as I think the footbridge and path might belong to a farm and be on private land. I can see the mouth of the track leading to the bridge but the Street View access stops 80 metres down the 500-metre-long track. I really hope there isn't a gate. Worst case scenario we can park on a lay-by on the road and make our way on foot. Even then we might not have access. I try to plan ahead and avoid too many unexpected situations like this one. When driving new roads, I like to assess the large roundabouts before approaching them; I have the same sandwiches for lunch, go to the same trainspotting locations. Knowing what something is going to look like or taste like brings comfort, particularly in situations that have a high potential to produce anxiety. Heading to a new trainspotting location to literally catch a speeding train, I feel sure it will be OK but I'm nervous about a farmer getting annoyed or us blocking their access. I need to push those fears aside. This is our best option.

Ryan bounds out from the station entrance and accelerates towards me, his backpack whiplashing against his body movements. I open the door for him, he throws himself into the passenger seat.

'Where is he? Where is he?'

I double-check Traksy. 'We really don't have long, mate, he's at Frome North Junction, about nine minutes from here.' I can see his headcode sitting in Frome North Junction on the loop north-east of the station. 'He's still there, Ryan.' I scroll down the line and see a late-running service

Class 59

Build date 1985–1995

Total produced 15

Number in service/preserved 15

Prime mover EMD 16-645E3C

Power output 3,300bhp

Maximum speed 75mph

Current operators Freightliner, GB Railfreight

Nicknames Shed

to Gloucester just passing through Blatchbridge Junction. 'He's stopped to let a late-running unit through, thank goodness.' We have some time. I check my mirrors, look over my shoulder and set off.

The drive to the bridge is very tense. Not a gate; a clear run in. We find a wider section of the access road and I pull the car into the brambles a little bit too far and have to clamber out on Ryan's side. We set up our cameras on the bridge, Ryan's towards the middle, mine more to the left as I want to get the train's side profile as it diverges towards Westbury. 'Gordon has just left the loop, Ryan.' The late-running service to Gloucester acts as our two-minute warning, passing through ahead of Gordon's arrival.

66524 banks around the corner, over a mile away; it's a dead straight track up to the junction beneath us. 'He's here!' I shout across to Ryan. My breathing quickens. It will take him around one and a half minutes to reach us, a nearly

---

Britain's Highest Station

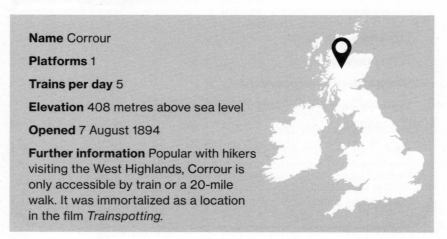

Name Corrour

Platforms 1

Trains per day 5

Elevation 408 metres above sea level

Opened 7 August 1894

Further information Popular with hikers visiting the West Highlands, Corrour is only accessible by train or a 20-mile walk. It was immortalized as a location in the film *Trainspotting*.

impossible length of time to withstand the suspense. I hear Ryan excitedly laugh and do a little hop. Gordon's headlights shine down the line on this overcast morning, approaching a section that cuts through a forest of thick trees. The puffs of different shades of green give way and bow down, as though Gordon himself has opened up the path. 66524 looks particularly yellow this morning – definitely seems like it has been in for a repaint – and the lights seem to be a new spec, incredibly bright. There is complete stillness in the air. The low pulsing of the EMD 710 power unit hasn't reached us yet and the train is still too far away for the tracks to have started hissing. 'Any second now,' I say to Ryan.

He turns and nods at me with a big smile on his face.

We wait. Gordon has to slow down on his approach to the junction; the diverging radius isn't that small so he will still have enough momentum to be carried into Westbury. Two tones are let off as he passes through the dense area of trees, barks that relieve the tension in the air, making me jump and chuckle.

A pause. Gordon coasts along. He's getting a lot closer and I'm worried that he hasn't seen us – I thought he might've done some more tones by this point.

'Maybe he thought we were at Pot Lane?' Ryan asks.

It's a possibility. I wave and give him a thumbs up. His luminous orange overalls glow in the cab and I see him reach for the horn. A high tone blasts from the top of the locomotive, *daaaaaa*. The low tone follows suit, but this time the high tone stays pressed while the low tone carries on, creating a harmony, *dAaAaAaAaA*. This sends lightning through my body. My laughing explodes into a scream. I have never heard such a noise on a railway, I didn't even know that kind of tone was possible on a class 66. I run behind Ryan to say thank you and wave to Gordon. 'Thank you, Gordon! Thank you so much, Gordon!' The tones themselves weren't that long, it was purely the harmony of the low and the high that set me off. I shout across to Ryan, choking on my laughter: 'That was mental!' Ryan is giggling.

I make sure I pan across with my camera, following the path of the locomotive for my video, careful to hold the camera stable, even though my body is jolting with emotion. Gordon passes under the bridge with fully loaded JNA wagons, the piles of aggregate completely homogeneous wagon to wagon. The wagons are loaded by a conveyor belt that drops the aggregate from above. While loading, the train itself moves forward at a predetermined speed, relative to the rate at which the aggregate comes off the conveyor belt, so the wagons are filled evenly and to the desired capacity, leaving a very satisfying geometric mound in each one. The wagons rush underneath. The red tail lamp flashes at us as the consist trails into Westbury Station.

Location: Over, 27 September 2021, 12:42

Long Marston is like purgatory for units, locomotives and carriages. They sit there in the snapping winds of Warwickshire, no longer maintained, awaiting their fate. Once a train has been seen through to its last day, it'll return to its train operating company's sidings where it will not enter service again. If left there, it would take up much-needed space that, in most cases, will need to be filled by its successor. Therefore, the retired rolling stock is taken to a long-term storage site such as Long Marston, dragged by a locomotive. That's if it's lucky. Some is sent at maximum line speed into the mechanical claws, turned to scrap just weeks after its withdrawal. Long-term storage is a much better option for a locomotive as it does at least offer a potential future. Carriages might be repurposed, and some firms may decide to invest in cheaper, extant rolling stock, instead of ordering some that is new. This is less often the case with passenger services, more with rail freight, particularly postal trains.

East Midlands Railway (EMR) retired their HST fleet on 15 May 2021. I had used their HSTs to ride up and down the Midland Mainline from Nottingham to London St Pancras, going to see my girlfriend in London or just riding on them for the sake of enjoyment. Spotting them was a treat too, one of my favourite encounters being in Normanton on Soar where you could hear the MTUs blasting through Loughborough, see them bank around the corner and fire under the bridge, thrashing right in my face. Leading up to the final weeks of service, EMR repainted two locomotives: 43274 in a new purple livery to match the rebranding of the company, and 43102 in the old Intercity Swallow livery, for good reason too. 43102 is particularly special as it holds the world record for the fastest diesel train. It was set on 1 November 1987, during a test run with 43102 on the front and 43159 on the rear. Between Northallerton and York they opened up the taps and reached a speed of 148.5mph. Unbeaten since. 43102 led the farewell tour from London St Pancras to Leeds, commemorative nameplate on the front, tones through all of the stations. I was lucky enough to be on that service up until Leicester. It was an emotional ride: tones echoing through Leicester on departure brought tears to my eyes. After 15 May, some powercars were saved, like 43102, which was put into the National Railway Museum. Most of the saved

Warning

Do not trespass
on the Railway
Penalty £1000

1 Oper
cross
or an

2 Cross

3 Close
after

powercars were purchased by heritage groups and freight operating companies. Others weren't so lucky. The remaining powercars and carriages were sent to multiple different storage sites, one of which was Long Marston. Today, 224 days since their retirement, the fate of ex-EMR powercars 43075 and 43061 is sealed. They are being dragged from Long Marston by 57312 to Newport Sims where they will be torn apart. We will be saying our last goodbyes.

The journey from Long Marston to Newport Sims will head up to Worcester Shrub Hill, where they will reverse and head back down the same line, splitting at Norton Junction, through Gloucester, out into Wales, finishing at Newport. Gloucester Station will be too busy for this sad moment.

'Where we going for this scrap move then?' Ryan asks as we cruise along the M4.

'How about Over?' I recommend.

A pause while Ryan thinks. 'Er, Over is bad, remember that farmer.' He purses his lips.

'We need somewhere picturesque and Over is perfect and close to Gloucester,' I say.

Ryan mulls it over. 'All right then.'

I'm surprised by his easy change of mind but I think usual limitations are going out of the window for this particular occasion.

This is the first time perfectly operational HSTs have been scrapped. No one wants them and it has become economically unviable to keep them stored. Thinking of how they could be flying down the line at 125mph over viaducts and through stations, ferrying passengers to and from the capital, makes me sad and I try not to think about it while I'm driving. Previous HSTs have been scrapped due to irreparable damage, for example after the Southall rail crash and the Ladbroke Grove rail crash. Comparatively, the images I've seen of 43075 and 43061 in Long Marston seem like they couldn't be more ready for another day in service. It could be argued that we need to move on and focus on new, more efficient trains. This is true, the environmental impact alone is huge, but it still feels like such a waste. I have such a strong root, buried deep into the ground, that connects me to my HSTs. I will say goodbye to every last one.

The Over crossings are just outside Gloucester, running parallel to the A40. My heart starts to beat hard as I slow. The kerb dips down and I pull the car in, crossing the shift of terrain on to rumpy rubble. The shrubs open out to a large circular turning area, the muddy brush carved into random slices by tractor tyres. Abruptly, I realize a locked steel gate bars our path through to the crossings. I was expecting to be able to drive a lot closer to them.

'Hang on, Ryan, one second.'

There's a sign on the fence crossing the path, impossible to read from here – and I have 20/20 vision! I turn the engine off, get out and stroll over to it. I crouch down but the writing is barely legible, Sharpie on a bit of plastic that has been zip-tied between the rails of the gate:

This is a turning zone ... DO NOT PARK HERE ... your vehicle will be removed

I turn back towards the car, making a 'not good' face to Ryan. 'We'll need to find somewhere else, mate,' I grumble as I get in. 'Is there anywhere near here?'

'Well, um, there's a McDonald's around the corner but it's a bit of a walk.'

We agree to park there. We have over an hour until the scrap move is due through so we can take our time.

Turning around in the tractor turning zone is nice and easy – it is a tractor turning zone after all and I'm in a car. Oddly, though, there are traffic lights leading from the rugged path back on to the road. It seems totally out of place, but on second thoughts, the A40 is quite busy and bringing a tractor plus trailer across two directions of double-lane traffic would certainly require the flow to be stopped.

The lights are red. We wait.

'Look at that Scania!' Ryan blurts.

A Scania R650 truck whooshes by, covered in LEDs and cool graphics. I whistle in appreciation. 'That was nice, V8 too. Did you see the exhaust pipes out the top?'

I share a passion for lorries with Ryan, which keeps us happy on our long motorway drives. On the way back from Stafford once, I saw bright LEDs brimming the edges of a lorry without its trailer, what seemed to be a Scania from behind. Overtaking it, I realized it was a Scania R730 V8. Bullbars, airbrushed paint job, chrome wheels and an open pipe exhaust that ran along the side of the lorry then curved out, chucking the exhaust just before the rear two wheel-sets. Both Ryan and I were bouncing in our seats. With my eyes on the road, I lowered the window on Ryan's side, the exhaust right next to us. It was burbling at an incredibly high volume: *whooo ho ho ho!* We were both loving it. The lorry then downshifted, the revs blipping, slashing through the wind noise from the open window, and accelerated. It was so insanely loud, Ryan was pumping his fist in the air; I was looking dead ahead but bellowing with laughter. A fantastic

Train Operating Companies

Operating Company	Region	Livery
East Midlands Railway	London–Midlands	Previous: white, yellow, orange, red, blue Current: purple
Greater Anglia	London–Anglia	White, grey, red
Merseyrail	Liverpool, Cheshire, Lancashire	Yellow
West Midlands Trains	West Midlands	Orange, purple, grey
London Overground	Greater London, Hertfordshire	White, orange, blue
Chiltern Railways	London–Buckinghamshire, Oxfordshire, Warwickshire, West Midlands	Previous: white and blue Current: white and grey
CrossCountry	West Midlands, East Midlands, South-west, Yorkshire and the Humber, North-east	Grey, pink, purple
Grand Central	London–Sunderland	Black and orange
London North Eastern Railway	Greater London–East of England, East Midlands, Yorkshire and the Humber, North-east, Scotland	White and red
Northern Trains	North-east, North-west, South Yorkshire and the Humber	White and purple
South Eastern	Greater London, Kent	Blue, navy blue
Avanti West Coast	Greater London, South-east, West Midlands, North-west, North Wales, Scotland	White, green, black, red

Operating Company	Region	Livery
Great Western Railway	London–South-west, South Wales, West Midlands	*Green and silver*
Hull Trains	London–Hull	*Navy*
Lumo	London–Edinburgh	*Blue*
South Western Railway	London–Surrey, Hampshire, Dorset, Berkshire	*Previous: white, yellow, orange, red* *Current: grey, dark grey, navy blue*
TransPennine Express	North-west, Yorkshire and the Humber, North-east, Scotland	*Blue, grey and purple*
Thameslink	South-east, Greater London, East Anglia	*White and light blue*
TfL Rail	Greater London, Berkshire, Buckinghamshire, Essex	*White, purple and black*
ScotRail Trains	Scotland, Cumbria	*Blue and grey*
Caledonian Sleeper	London–Scotland	*Dark green and silver*
Transport for Wales	Wales, North-west, West Midlands, Gloucestershire	*Light grey and red*
Southern	London, Sussex, Surrey, Hampshire, Kent, Hertfordshire, Buckinghamshire, Bedfordshire	*White, light green and dark green*

memory. It makes me feel happy even thinking about it now.

The light is still red.

'It's not changing, is it?' Ryan laughs.

I try reversing back and forth to try and get the light to change. They must operate on a sensor as tractors must use this area so infrequently that an alternating traffic system with a heavy bias to the A40 traffic would still be inefficient. Sensing movement from the path should trigger the lights to turn red for the traffic cutting across. The A40 traffic is presumably sensor-based as well because it must select the best time for the lights to turn red to maximize traffic flow. I love visualizing flow systems and how they can be impacted by certain conditions. I even watch traffic flow optimization videos on YouTube. I find the structured order of the networks particularly satisfying. My brain gets an equal buzz from building dams in rivers and redirecting small sections of water. I could do it for hours.

Anyway, the light remains red and we're going nowhere. I still think it's because the traffic light hasn't sensed us.

'Ryan, can you wave at the traffic light, please?' I open his window and he manically starts to wave at the taunting red light. 'Keep going, please.' But nothing is working. Staring into the redness of the secondary traffic light across the road, we wait and wait and wait.

'Oh shit.' Ryan is looking over to the right and pointing. A towering tractor with ploughing equipment on the back is bouncing down the road on tyres as tall as me. 'No, no, no, that's not good.' Ryan's voice is trembling.

My adrenalin starts to rush but in a different way from when I'm with my trains. The worry makes my heart beat a lot harder and slower than when I'm happy, thumping in my chest, making it hurt a little bit. When I'm getting a buzz from a train, my heart bounds along; the beating feels lighter and quicker. I pay attention to this because I have a heart condition that is caused by inflammation of the heart lining, the pericardium. It's caused by stress and is something I need to keep an eye on. On first onset in 2020, my heart hurt so much I could only walk at a snail's pace. I'm a lot better now, but stress and anxiety still cause me pain.

'He's going to want to turn in here,' Ryan says.

I quickly adjust my car so we are as far to the left of the clearing as possible. Still not enough room for it. I consider reversing into the corner of the turning area but the farmer is still going to want to use one of the gates and the ploughing equipment is so wide that I'd become trapped. 'I don't know what to do, Ryan.'

The tractor approaches. I can see the farmer bouncing around on his air ride seat. He has a green light. I watch him for any angry facial expressions or directions telling me what to do. He stares at us. The tractor drives by and continues along the A40. Phew.

Over Farm Shop turned out to be our best parking option. The traffic light turned green eventually, and I found the shop just twenty seconds after departing our gravel trap. Ryan wasn't so keen and I'm pretty sure he knew about it before. 'I bet they don't like you parking in here,' he murmured as we pulled in.

'It's fine, mate, let's go in and buy some drinks or something.'

Clutching two cans of San Pellegrino, aranciata flavour, we are now making our way across the A40 and on to the pavement on the other side. Strolling along, I can feel the cortisol that was in my system earlier being broken down and expelled. Cars and lorries rushing by, we keep looking over our shoulders to see if there is anything decent coming. A few Volvos, a Mercedes and two Scanias give us toots or flashes as we wave at them. An apple tree hanging over the pavement has dropped all its apples on to the ground. Ryan, not looking down, accidentally kicks one which knocks into another, both of them rolling into the road. 'Ryan!' My buttocks clench as I have visions of a passing motorbike having a banana-skin moment. The first apple rolls under a Nissan Qashqai in the first lane and is then completely flattened by a Honda CR-V in the outer lane. The second apple rolls after it, does a funny hop and rolls back into the gutter at the side of the road. We pick up our walking pace, trying to lose the embarrassment and avoid any chance of getting told off, even though it was an accident.

There are three different crossings at Over. We pass the gate with the scrawled sign, following the public footpath arrows. The first crossing is only 20 metres after the gate, with a good view towards Gloucester and Over Junction

Crossing Types

Level crossing	For all road vehicles, which must obey warning lights and barriers signifying an approaching train.
Foot crossing	For use by pedestrians, who must follow instructions on a sign; in some cases a signaller will need to be phoned before proceeding.
Farmer's crossing	For use by farmers, and in some cases pedestrians, if a public footpath follows the route.

Signal to the left; to the right the track stays straight for around 125 metres then becomes concealed by trees as it gradually cambers to the right. Behind us, facing the path we have just come from, there is a field that seems to have just been harvested, stalks poking up from the ground, sheared off by machinery. The path immediately turns 90 degrees to the right and follows along, parallel to the tracks. Ryan recommends we head to the third crossing as the view is apparently a lot nicer – a twelve-minute walk. We have twenty-five minutes until the scrap move passes by.

Birds of prey hover overhead at the third crossing: there is a pond over the other side where the male ducks are burbling and the females are producing more distinct short quacks. The view is certainly a lot nicer here. Looking over the tracks, past the pond, the horizon extends all the way to the Cotswolds. Gangs of Lombardy poplars sway above, rustling intensely when the wind blows through. At the crossing, the path takes another 90-degree turn to the right, down to the farm where large sheds and farm machinery lie still. If it wasn't for the public footpath sign I would've certainly steered clear to avoid getting told off by the farmer. As I have heard from Ryan, this has happened before, but he apparently just seemed irritated that Ryan was trainspotting, loitering instead of walking. To the right, the line straightens out and dips down, heading under a bridge towards Lydney and deeper into Wales. To the left, the track has a fantastic camber to it, picking up from the end of the corner seen at the first crossing. It's uphill too, until the crossing, so it should almost give the sensation that the train is in flight, soaring upwards like the birds of prey above us. The angle of ascent and camber isn't as steep as a bird would experience in flight but the nature of a train, being so huge and carrying such momentum, makes even the slightest offset from the horizontal plane seem drastic.

We are standing near the fence, waving at the occasional CrossCountry class 170, when, from behind us, we hear the distinctive sound of a tractor. I hesitate to look.

'Oh no, not him, please,' Ryan says under his breath.

'Don't look, Ryan, let's focus on the trains.'

The whirring comes closer, the rpm of the engine bouncing as it passes over rough terrain.

'What're we gonna say? I don't know what to do.'

Ryan looks upset, which gives me a bit of confidence to try and take charge of the situation. 'Don't worry, mate, I'll do the talking. Don't worry about it.'

The tractor comes closer. He's certainly here for us. It slows down, stops, the engine returns to idle. My heart is thumping hard in my chest again. I can't help but look as the driver slams his door shut and jumps down. A man in his fifties with a cap on. I smile, trying to come across positively.

'Anything good coming through?' he asks.

I'm stunned. 'Erm, yep, there's a class 57 dragging two class 43s to the scrapyard.'

'No steam then?' he replies.

'Nope.' I shake my head.

Ryan turns round, looks at the farmer, then looks at me, puzzled.

'All right then, enjoy, chaps.' And with that the farmer walks off, hops back onto his tractor and continues down the path.

I look at Ryan. I feel so relieved.

'Different farmer,' Ryan remarks, knowingly.

57312 dragging 43075 and 43061 is just passing through

the centre line at Gloucester Station. I'm looking down the track but have to step away for a moment because I can feel myself starting to choke up. I can't help but think about the times I blasted up and down the Midland Mainline behind these locomotives, feeling the sensation of speed in my body, hearing the clattering of the wheels over the points and coaches whining at maximum line speed. The MTU power-cars will not be turned on for this scrap run, they will be dead in fact, but the rumble of the MTU powercars at full chat is running through my mind; they used to thump the air through Loughborough like nothing else. I know them so well, they are so familiar. They offer me comfort.

Class 57

Build date 1998–2004

Total produced 33

Number in service/preserved 33

Prime mover EMD 12-645, 57/0 & 57601: 12-645E3, 57/3 & 57602–57605: 12-645F3B

Power output 2,300hp

Maximum speed 57/0: 75mph
57/3 & 57/6: 95mph

Current operators Direct Rail Services, Great Western Railway, Locomotive Services Limited, Rail Operations Group, West Coast Railways

Nicknames Body snatcher

My mind strays. I think about other things I have lost recently – my cat Macca Pacca, Macci for short, who was my best friend. He passed away six months ago. He used to understand exactly what I was feeling and loved curling up and being cradled like a baby in my arms. In many ways he and his brother, who is still alive and is a lovely boy, helped in terms of companionship when I moved to Somerset. We moved in 2006, we got them in 2007. Macci saw me through primary school, miaowing for food when I got home. He was there for me when I felt lonely through secondary school. He saw me leave home and go to university. He was there waiting when I came back during lockdown. In many ways Macci and trains had much in common, giving me a sense of comradeship, comfort, inner peace. Saying goodbye to Macci was incredibly hard because I had lost my best friend. Saying goodbye to class 43 HSTs isn't as sad, but here and now it is

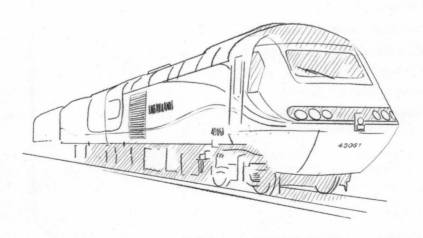

still making me cry. There is a deep sense of loss in saying farewell to something that I've known and loved nearly all of my life.

57312 appears from around the corner, coasting along. The tears are rolling down my face. The driver gives us a single, mournful tone. Track hisses, the ducks quack and the trees rustle. I let out a sob. The locomotive passes us. A bar attached to the rear of the 57 leads out and into the mouth of 43075 as if it were a fish hooked on a line, being pulled out from the water. The front of 43075, what should be the leading edge of a train that cuts through the air, is instead shadowed behind the class 57 that's merely completing its duty. 43061 is attached directly to the back of 43075. Where there would've been coaches full of happy passengers cruising through the countryside, there's now only a sad, sagging gangway that locks the pair together. Holding hands until the end.

'Goodbye 43075, goodbye 43061,' I whisper from behind the fence.

43061's face is our last sight of the train moving off into the distance. For a moment, with an MTU rumble, I imagine nine carriages are between both powercars, out on an excursion into Wales with heat haze rolling off the back, distorting the face I know so well.

Chapter Seven →

Chasing Steam

Location: Frome, 5 March 2022, 04:48

A symphony of strings grows in volume until the puny speakers in my phone are maxed out. I rise from my sleep in a bit of a daze – I was in the middle of a deep dream. I am staying at my parents' house in Somerset. It's a familiar IKEA mattress, nice and firm, just how I like it. I press stop on my phone alarm. I chose to change my alarm from the typical 'radar' noise, as any time I heard it during the day, it'd give me a shock. The 'dawn chorus' alarm is a little better but ends up sounding like a screaming child if left for too long. It's twelve minutes to five. There's a railtour departing from Westbury at 05:12. I swing my legs off the edge of the bed and have a moment of contemplation, torn between staying here in the cocoon of warmth or breaking out from the crust and kickstarting my day. I know that I will regret it if I don't go so I fling on my clothes, grab my GoPro, wallet and keys, and plod out of the room and out the door as quietly as I can.

My Mercedes is back from the garage. I am glad as the heater gets to work a bit quicker than Lucy's. I have a long day of driving ahead, seeing two different railtours: the

Pathfinders tour from Westbury to Crewe (Pathfinders Rail-tours is a company well known for putting vintage traction together with some mark 1 carriages and running routes infrequently used by passenger trains) and the *Flying Scotsman* from London Paddington to Worcester Shrub Hill. The main event will be the *Flying Scotsman*, a locomotive with such history. It has travelled the world, blasted up and down the East Coast Mainline and become the poster locomotive for train enthusiasts around the globe. I just need to make sure I get some good footage of it. Last year I caught it heading through Gloucester with Ryan. It was a sweleteringly hot day and I was on a crowded bridge packed with babies, toddlers, children, teenagers, adults and the elderly. I had to stand on my fold-out camping chair to try and get a good shot of it. Unfortunately I only got the front profile as it was coasting downhill under the footbridge. I was blasted in the face by

A3 *Flying Scotsman*

Build date February 1923

Number in service/preserved Retired

Power output 1,800–2,000hp

Maximum speed 100mph

Current operators London and North Eastern Railway, British Railways

Nicknames Flying Moneypit, Flying Scrotum

soot, which made both Ryan and me laugh after this slightly underwhelming encounter with the legend. I'm hoping for a side-profile shot with it carrying some speed and also a shot of it pulling away from a station. I think the best bet will be to go to Didcot Parkway for the pulling-away shot with Worcester Shrub Hill as a back-up, and to catch the at-speed shot at Ascott-under-Wychwood Station. I've checked the schedules and I know these will be the best options for the outbound journey.

Mine seems to be the only car on the road this morning. It's completely dark, there are light patches of mist that don't seem to faze her headlights, and the effective cat's eyes are few and far between. The width of the road varies. Sometimes the centre line disappears altogether – a clear reminder to stay focused in case an oncoming driver appears in the centre of the road. My eyelids are locked in the 100 per cent open position, barely blinking, the heaviness of my eye bags weighing down my lower eyelids. Streetlights greet me as I pass under the rail bridge that leads services through Dilton Marsh and down to Westbury.

Location: Westbury Station, 5 March 2022, 05:08

Westbury Station is completely empty. There is only one train at the platforms, the railtour to Crewe. The bright red DB Cargo 66084 is singing away with its very distinctive idle – *ying ying ying ying ying*. It's a sound heard frequently on the railways in the UK. The power unit used is reliable, powerful and relatively kind to the environment. The class 66 adopted this power unit and has since been ordered, reordered and reordered again after proving itself a trustworthy workhorse for our rails. With each revolution of the EMD 710 power unit it lets out a little chime or *ying*. The reason for this unusual sound is that the engine is a two-stroke. Unlike a more common four-stroke engine, where there's a process of suck, squeeze, bang, blow, the exhaust cannot rely on being pumped through, so the inlet air needs to be at a high pressure to push the exhaust gases out when both the inlet and outlet/exhaust ports are open. At high rpm the turbocharger is spinning so fast that the air it pushes in is at a high enough pressure to expel the exhaust gases from the combustion chamber. At low rpm the turbocharger is not spinning fast enough to generate enough pressure so the revolution from the drive shaft aids the turbocharger in blowing air into the combustion chamber and thus pushing out the exhaust. It's this mechanical assistance that causes the *ying ying* noise.

The class 66 will be hauling a rake of cream and brown mark 1 and mark 2 carriages, traversing routes typically reserved for freight trains. This is particularly interesting for train enthusiasts who collect mileage or who are looking to tick off the lengths of all the tendrils on the network. I love traversing new lines but I am more interested in following this railtour to its third stop, Cheltenham Spa.

There are only two other trainspotters at the station, both with tripods on the end of platform 2. The train itself is more empty than I expected. Not many passengers have joined here at Westbury; I imagine the bulk of them will jump on at Bristol or Birmingham. It has to leave here at this time in order to potter its way up to Crewe via all the particular loops, sidings and freight lines that make the tour so attractive to us enthusiasts. The carriage interiors are illuminated by a much warmer light than in modern-day carriages, brighter too. It seems a little eerie. It's almost as though the ghost train has pulled in, collecting train enthusiasts from the past and bringing them into the day of unusual routes and freight sidings.

A driver from DB Cargo potters up the platform towards the locomotive; he seems to be in his late sixties with light hair and a sweet smile. I walk a couple of metres behind him. He opens the thick steel door of the cab of 66084 and hops in. Getting closer to the locomotive, I feel the mechanical heartbeat through my body. The sound of the blower going *ying ying ying* is now paired with a throbbing undertone that resonates with part of the outer shell of the class 66, creating a rhythmic rattle. The exhaust emerging from the roof catches the platform lights but, unlike a class 37 for example, this is the only exciting thing that appears from the top. With a class 37 or any English Electric or Sulzer engine from that era, there's very little silencing of the noise coming from the exhaust, basically allowing thunder to exit straight from the engine. The class 66s on the other hand are thoughtful, modern locomotives with a whole load of muffling sitting between the combustion in the cylinders and the atmosphere above it. Not quite as sonically satisfying, but the power of the locomotive is palpable and it can still tap into my nervous system when the engine thrashes.

The driver turns on his cab light, and after setting up I see him sitting and waiting for departure. He has five minutes and I can see the doors of the carriages are still open. I feel that it's a good opportunity to go and speak to him.

'I think I'm going to try and beat you to Cheltenham,' I say with a cheeky smile.

The driver leans out from the cab door. 'Oh really, how're you going to do that?'

'I have my car!'

'Ahhh right, I'll see you there then maybe.' He smiles and winks before shutting the cab door and returning to his driving position.

A few minutes later, the doors are closed and the signal turns green. The driver gets the all clear from the guard and he puts the 3,000hp locomotive into notch 3. The *yinging* increases in frequency. I imagine in my mind that the locomotive has a mouth and in the idle position its jaw is moving up and down at the rate of *yings*, with a calm face; upping the notches, the mouth tries to keep up and its eyes widen until, at the higher notch range, its mouth is just wide open, no longer *yinging* but shouting, facial expression akin to an angry bear. The driver moves up to notch 5, chucking up a load of clag from the exhaust as it passes me. I can't help but screw up my face at the sound as the EMD 710 power unit roars. Passing under the signal, the orientation turns immediately from green to red, casting over the top of the cream and brown coaches. They pass over the points under the roadbridge after the station and, leading the way, 66084 hauls the vintage carriages in the direction of Bradford-on-Avon, then Bath Spa. The lights from the interior pan across the nothingness of the early morning, glowing on the ballast, track and foliage that surround the route out from the station. Red flashes from a lamp at the rear carriage highlight the end of the rake of coaches.

The station returns to stillness and the two other train-spotters head back towards the exit. I give them a smiley nod and they nod back. I get a fast walk on – no point dawdling around if I'm trying to beat the train to Cheltenham.

Oh my goodness, I need to pick up Ryan from Gloucester on the way! I totally forgot that I'm making a quick stopover. Luckily Gloucester is kind of en route. I just need to dip off the M5 and jump back on again. The only issue is that it's a bit of a maze to get to Ryan's house. I check Google Maps, factoring in this extra time, and it still seems like I can make it. I hop into my car, turn over the still-warm engine, do a U-turn and make my way to Gloucester.

There's a pothole I always have to negotiate when turning on to Ryan's road. I swing around a little more than usual, successfully dodging it. He's there, standing next to a space for me to pull into. He opens the door and I quickly inform him, 'Ryan, we really need to hurry, the 66 tour is passing through Filton Abbey Wood now!'

'Bloody hell, Francis,' Ryan replies. 'You took your time, didn't you?'

I don't take any time to reply to his comment as I'm already checking my mirrors to pull out on to the road again. The sun is starting to rise, the clouds are showing hints of pink candy floss. As we pull on to the A40, we are chatting away about the day and the plan to head to Didcot Parkway after Cheltenham Spa. To my right, the clouds have exploded into peachy salmon skin: the high-altitude cirrocumulus clouds look like fish scales, or the fur pattern of a tabby cat. I remember that when one of Granny's tabby cats, Basil, died, probably around 2005, my mum pointed to the sky, when the clouds were very similar to the way they are now, and said, 'Look, Basil is in the sky.' Ever since then I have thought about that every time the clouds were cirrocumulusy. I point the sunrise out to Ryan and say that it's a sign that today is going to be a good day. He nods in agreement. We start talking about how CrossCountry are scaling back their use of their class 43 fleet, which takes us all the way to Cheltenham Spa.

Britain's Lowest Station

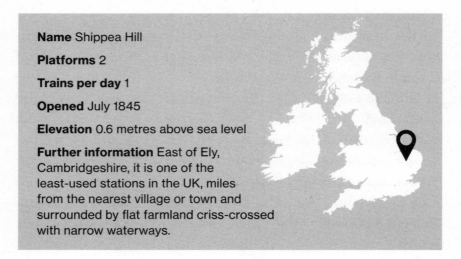

Name Shippea Hill

Platforms 2

Trains per day 1

Opened July 1845

Elevation 0.6 metres above sea level

Further information East of Ely, Cambridgeshire, it is one of the least-used stations in the UK, miles from the nearest village or town and surrounded by flat farmland criss-crossed with narrow waterways.

Location: Cheltenham Spa Station, 5 March 2022, 06:41

As soon as we arrive, I check Traksy. 'Phew, it must've been held somewhere along the way, it's four minutes late.' We jump out and I quickly pay for parking. Ryan runs ahead down to the platform; luckily the gates are open because it's so early in the morning. There are only a few passengers dotted around, waiting for services either to Bristol or Birmingham, I imagine. We head up to the north end of the platform, next to the roadbridge. There's a video on YouTube that was filmed here where a class 43 in Virgin XC livery powers out from the station with its Paxman Valenta engine screaming to the high heavens and chucking out loads of clag. 'Those were the good old days,' I say to Ryan as I remind him of it.

The railtour is only two minutes away. We establish our positions on the freshly resurfaced platform, the yellow line crisp and the Zone 1 indicator surrounded by orange satisfyingly sitting on top of the tarmac like icing. A GWR class 800 pulls into platform 1, on the other side.

Cheltenham Spa has a curved platform so I can see when 66084 is coming, its high-intensity LED lights reflecting off the side of the class 800, twinkling and signifying its imminent arrival. The class 800 is on diesel mode due to no overhead lines providing electricity, with the five engines underneath whirring away. One of them right next to us is cheekily polluting the platform with modern, muffled diesel noise with a whiny overtone, slightly frustrating me that I haven't been able to tune into the EMD 710 power unit of the class 66 yet. The class 800's power units idle at a higher rpm, due to their being comparatively smaller – *dugadugadugadugaduga* grumbles from underneath the carriages with a modern sounding electrical overtone. The class 66's idle note starts to increase in the soundscape as it approaches, the *ying*ing oddly harmonizing with the overtone of the class 800. The freshly painted yellow and red front peeks around the corner, the coaches follow, and screeching starts to echo around the station. Class 66s have very handsome fronts, subtly angular from their side profiles with shallow pointed rooflines, giving them their nicknames – sheds. As 66084 draws to a halt, the irritant of the whirring class 800 disappears and the class 66's idle sound sits right next to us in its purity.

The driver's window opens. 'You made it!' He points at me with a smile across his face.

'I beat you!' I respond cheekily.

He glances down at his watch. 'Only just.'

I turn to Ryan, and there's a smile across his face.

A few passengers are helped onboard and the mark 1 coaches are shut by the stewards with satisfying clunks. Brake pressure is released, signifying departure in the next ten seconds. The *yings* increase in frequency. I can tell the driver is going to take it up to the higher notches as he has a clear line ahead with no diverging points. Up to notch 3, the train starts to move. Each carriage has a small amount of slack between it and the next one as they bunch up under braking; pulling away causes the slack to be taken up coach by coach, sending a small shockwave right to the rear carriage, which then reciprocates and reflects back towards the front, a lot more subtly but you can just about hear it. Now up to notch 5, the locomotive is hauling all the coaches under tension. The red shell of 66084 ripples with energy as the 3,000 horses start to take traction, 12,000 hooves gradually extending, careful not to slip on the smooth railhead. My body tingles, feeling the vibration in the air. *BAAAAP BAERRR!* The driver lets off two powerful tones as he approaches the bridge; they reflect off the retaining wall and back into the station. Both Ryan and I jump and burst into laughter. The tones echo in the tunnel as the class 66 passes through. We are still laughing as I consider whether he did the tones for me and Ryan or if there were track workers on the line ahead and he was warning of his approach. We wave at some of the passengers, who don't have time to respond due to the increasing speed, but I hope it brings them some positive feeling for their journey ahead. We bid farewell to the class 66 railtour as it's about time we set off to catch the *Flying Scotsman*.

Cheltenham to Didcot Parkway takes an hour and twenty minutes: head south-east to Swindon, along the M4, eastbound, then up to Didcot. The journey by road is far less picturesque than the train journey would've been, meandering up arguably one of the prettiest routes in the western region, the Stroud Valley. Right at the top of the winding incline is Sapperton Tunnel, the same spot where we were able to see those class 37s coming from Lydney.

Location: Didcot Parkway Station, 5 March 2022, 08:11

The twenty-minute pick-up and drop-off spaces at Didcot Parkway Station car park will not be sufficient for our catch, and there are ANPR cameras here too waiting to catch anyone who overstays their welcome. I bite the bullet and pay £7.20 for a day's parking, which is the minimum amount of time you can spend in the car park if you are paying. I feel like there should be more parking options for trainspotters who may only want to be at the station for an hour or two, maybe less. I guess they wouldn't make as much money from us then, though, so where's the incentive?

The *Flying Scotsman* is due at 08:27. Passing through the entrance, I assume the platform will be packed. The gateline staff even know about the special service and don't hesitate to let me and Ryan in. We climb the stairs leading up to the platform. Very few people are actually here. We are both shocked. All in all, I'd say there are about fifty people waiting to see it. I thought it'd be in the hundreds! I suppose it's Saturday and people are happy to finally have a bit of a lie-in; they can always catch it in the evening on its return journey.

It's breathtakingly cold. I haven't worn my scarf today, which is a big mistake. Ryan is blowing his cheeks out in protest at the temperature. 'It's freezing,' he says, hopping from his left to right foot seemingly to warm up.

I can see a GWR class 800 approaching. An announcement projects through every speaker on platform 1: 'The next train on platform 1 does not stop here, please stand well clear of the edge of platform 1.' It certainly isn't stopping. The tracks sound like they're exhaling as the vibrations of the wheels travel through the air, producing an other-worldly sound, *waaaaaAAAHHHH*. Tearing past, it rockets through the platform at 125mph, the fastest any train can run in the UK (outside of HS1, from London to the Channel Tunnel). The gap between each carriage causes a change in air pressure

Class 800

Build date 2014–2018

Total produced 80

Number in service/preserved 80

Prime mover MTU 12V 1600 R80L

Power output 750hp per unit

Maximum speed 124mph

Current operators Great Western Railway, London North Eastern Railway

Nicknames Cucumber

surrounding the train – *phwoaow phwoaow phwoaow*. The rear follows almost as soon as the front has passed through. 'Used to be better when it was an HST,' Ryan remarks. He's right. Just like at Cheltenham, the station used to be frequented by the class 43 HST, arguably the train that saved British Rail, due to its journey-time-shrinking capabilities and percentage-availability improvements over its predecessors, nearly fifty years old, still running on lines around the UK.

Three lamps, the centre one high-intensity and the outer two slightly dimmer, sit across the front of the *Flying Scotsman*; they peer straight down the line, eyes focused. The bustling on the platform suddenly reduces; cameras are drawn and positions are established. It takes a little while for the train to make its way to the platform.

I don't find steam as exhilarating, I have to say – it's a controversial opinion, but it is mainly to do with the sound. Steam doesn't course through my veins the same way internal combustion does. Steam traction is satisfying in a different way though, the massive cast wheels, beaten panels and sturdy con-rods all harmoniously puffing and shifting together. Converting the intense capsule of energy in the boiler into linear motion is truly beautiful, and that is what attracts me to steam: it satisfies my engineering brain. Not so much a nervous-system tingler though.

She coasts in, knocking, puffing and clunking. We follow her down to the end of the platform. All the train enthusiasts have come from their initial locations and converged on the beast at the end of platform 3, like bees returning to a hive. She sits there releasing pressure, shooting out steam from the top of the boiler with tremendous velocity, peaking my eardrums to the point that I can only hear buzzing. The men on the footplate are sooty, some old hands, others youthful lads; all seem like powerful uber humans, superheroes, taming the beast with careful inputs. Her wheels, rods and pistons are freshly greased and oiled, now motionless but ready to move in an instant. We are all crowded around the locomotive, jostling. There are a few grumbles of passive-aggressive comments – 'Oh, excuse me' – for we are all trying to get a good shot.

The *Flying Scotsman* settles in Didcot Parkway Station for four minutes after blowing her steam and with all doors shut. A short blip from the whistle and a nod from the driver, and the pressure starts to push the pistons. *Puff* – a gentle release. Briefly the force from the steam overcomes the friction of the rails and she stumbles, wheels slipping on the tracks, puffing at the same rate as the rotation of the wheels, momentarily sounding like she's travelling at around 30mph. Regaining traction, she gathers herself and gracefully pulls out from the station, coaches following. *Puff puff puff* – plumes of enveloping steam jettison out from the top of the locomotive. It clatters across one set of points, then another, proceeds around the corner and then really opens up the taps.

As the puffing continues, I turn to Ryan. We both share the same facial expression: a little bit miffed. I'm happy with the encounter but I was a bit dissatisfied by the arrival and departure. I couldn't get the right shot and the cold didn't help either. Ryan agrees with me: 'I couldn't even get my camera over the bloke in front of me – he was well grumpy.' We turn around and head for the car park. The next station we plan to go to, she will be blasting through at speed. Hopefully we can get a good shot there.

Location: Ascott-under-Wychwood Station, 5 March 2022, 09:41

We arrive forty minutes before the *Flying Scotsman* is due to pass through. Ascott-under-Wychwood's car park dips and undulates, my car rocking in all three axes. No tarmac here, just dark brown mud, puddles and the occasional patch of gravel emerging from the compacted surface. I have to put the nose of my car in a bush in order to do a three-point turn. The surrounding cars have not left much room so I have to squeeze into a semi-space between a blue Ford Transit and a Skoda Yeti. The car park is small, probably around twelve spaces, two of which are for Network Rail vehicles; I imagine later on in the spring, the spaces might reduce further owing to a bolshy-looking buddleia. You can see on to the platform from the car park, slightly above foot level. There are only two spotters at the moment, both with tripods, and they seem to be veterans.

Ryan and I make our way up the disabled access ramp to the platform. It's my first time at the station – very pleasant, a level crossing at one end with a fantastic signal box which

almost looks like it could still be in use. The line used to be single track, updated in the early 2010s with platform 2 being installed then. Yellow lines on platforms are an important indicator of the speed at which trains can pass through the station. Here, the line encroaches on around two-thirds of the narrow platform space, which causes a bit of discomfort for me. Restricted-width platforms mean limited space for getting the right shot and you have to stand with your back practically against the fence when a fast train approaches, which is probably why there is an open-to-the-elements shelter that sits in a recess, completely stepped back from the platform.

Ryan suddenly turns round and runs back in the other direction. 'Francis, I forgot my tripod, can you open the car, please?' I sarcastically roll my eyes – 'There you go, Ryan.' One click of the key and it unlocks, allowing Ryan to retrieve his tripod from the boot.

In both directions it is clear, a good 500 metres one way until a gentle, declining curve to the right, and in the other it's dead straight for over a mile. The line speed is 70mph so the *Flying Scotsman* will be gunning for it along here.

'Francis, look what I found!'

I turn and find Ryan holding up a smashed iPad.

'Looks good, don't it?'

I nod in agreement. 'Could take it to CEX and see how much it's worth,' I joke.

Ryan paces down to the on ramp, doubles back and re-establishes his position. A crowd has started to gather on the platform. I suspect there will be more people coming.

Fifteen minutes later, the platform is rammed. Families, elderly couples, trainspotters. All out to see the legendary locomotive. I could put money on there having been a post on a village WhatsApp or Facebook group: 'The *Flying Scotsman* is coming to town!' I worry slightly about the children near the edge of the platform; parents are keeping a watchful eye on them but I am still slightly on edge. There's an old lady next to me who I can tell will end up getting in the way of the shot if she stands a little more to the left. I don't mind too much as I can always shuffle over, and ultimately, I want people's experience of the *Flying Scotsman* to be as great as it can be. I've been told to get out of the way of shots before and it has burst my excitement bubble slightly, so I wouldn't want to do that to the old lady.

A whistle in the distance cries towards the platform. Both seasoned and first-time trainspotters move to try and see over the person's shoulders in front. It's coming at some speed. Doesn't seem to be 70mph, more like 55ish – still

impressive for a steam locomotive. I don't see her until she's halfway down the platform – too late. She lets off a tremendous whistle though, and the lady in front of me jumps to the right. Two seconds of seeing the face of the boiler, the muscular shoulder, pistons and rods shunting back and forth, the footplate with the same chap who nodded at me at Didcot, the tender that smoothly marries up to the cross-sectional area of the carriages, the steam that is pushed down along the top of the coaches by the air. *Whoosh whoosh whoosh* – the coaches with faces keenly looking through the glass, eyes locked with mine for a fraction of a second. Passing along in a flash. The red tail light blinks into the distance, now a companion for my disappointment. I was really hoping for a clear shot and I was too worried about the people on the platform to completely enjoy the moment. She dips off to the right and continues on her way to Worcester Shrub Hill.

Ryan and I are mostly silent on our way to Shrub Hill. I'm quite tired after my 04:48 wake-up and there's a low mood after the two mediocre encounters. Worcester has three different stations: Worcester Shrub Hill, in the east of the city, opened in 1850, dark grey brick, wide, with white accents on its facade; Worcester Foregate Street, more central, opened ten years after Shrub Hill, relatively facadeless, neatly nested within the city; and Worcestershire Parkway, opened in 2020, which is a modern intersection of the Cotswold Line and the line heading from Bristol to Birmingham, allowing swift changeovers or a fifteen-minute drive into Worcester city centre. We are heading to Worcester Shrub Hill, the last station on the *Flying Scotsman*'s journey today. Shrub Hill will provide a perfect environment for the locomotive to settle down in: a lot of original features, quaint waiting rooms, signalling by wire and original clocks.

Location: Worcester Shrub Hill Station, 5 March 2022, 12:09

Vehicles are parked everywhere. The road leading up to Shrub Hill is lined with car after car, I assume all here for the *Flying Scotsman*. I catch a glimpse into the station. 'Why are there security and barriers?' I think to myself.

'Oh my word, Ryan, it's packed!'

I stop the car at the entrance and can see through the doors that the nearside platform is bustling with people. 'Ryan, it's already there!' The burgundy coaches poke out from behind the mass of heads moving around. 'Yes, look!' I point to the left of the station, where steam is being chucked into the air. 'We need to be super quick, Ryan.'

Fortuitously, a Kia Soul edges out from a space right next to us. I swing in behind it and we jump out of the car. We scale the footbridge two steps at a time, but it's not enough. The *Flying Scotsman* lets out a toot, chucks clouds of steam into the air, and hauls its coaches into the sidings where it will rest until its return journey. Ryan and I stand on the bridge with our hands on our heads, both of us really disappointed.

We get back in the car, doors shut. We sit there and sigh. We were counting on the outbound journey for getting a good video, with far better light than there will be this evening. The cloud is low and thick now. Any of the good spots on its return journey will be shrouded in darkness by the time it gets there. We mull over what to do.

'Probably best to go to McDonald's, don't you think, Ryan?'

He nods in silent agreement. I can tell he's just as disappointed as I am.

Sapperton Bank would have been the perfect location to film the *Flying Scotsman* if it was running three hours earlier and if it wasn't cloudy. A good few miles of pure incline, weaving back and forth until plateauing into the mouth of Sapperton Tunnel. She will have to work hard getting up there and it will sound fantastic – which is the main reason I'm persuading Ryan to go there. Even if it's dark, we will be able to hear the reverberations of the locomotive around the valley. I'm very curious to see what it sounds like powering up the bank.

'All right, that's fine, Francis, but do you mind dropping me back in Gloucester after, please?'

Ryan has agreed to Sapperton Tunnel. After all, we've had a special moment there before with the class 37s.

Location: Sapperton Tunnel, Gloucestershire, 5 March 2022, 16:25

The last time we were at Sapperton Tunnel, we realized, after traipsing through miles of country lanes and whacking down bushes, that we could've parked the car only 50 metres away. Remembering this, we pull up in a lay-by right opposite the gate that leads down to the tunnel, left wheels entrenched in leafy mud, right wheels unscathed. We find the opening in the bushes. The valley opens out in front of us, the view we now know so well after repeatedly watching back the class 37 video we filmed here last time. The bushes surrounding the track have been cut back slightly, the air is drier and the clouds are a lot higher this time, though still thick and light-restricting. It's nearly half past four. The

Flying Scotsman is due through at 17:49.

There's a man here with his grandchildren, sitting on a camping chair, one child on either knee. Two Staffordshire bull terriers are stationed either side of the chair. They are quite plump and are panting, even though they seem to just be sitting there. They aren't intimidating. I smile with pursed lips and he nods back.

'Oorite, chaps!'

I turn, look up and see a man standing behind me nestled in the bushes a couple of metres above the natural terrace we are standing on, presumably for a better shot at a higher viewpoint. He has his tripod set up in the undergrowth, legs splayed linearly and purposefully through the chaotic mass of branches. 'Let's hope these clouds bugger off,' he says as he peers through the viewfinder of his gigantic camcorder, lining up his shot. Ryan and I murmur in agreement.

A few GWR class 800s pass in either direction, blasting tones at the whistle boards up and down the valley, to alert any amblers to an approaching train. Miraculously, the clouds on the horizon have started to dissipate, or rather the sheet of low-level cloud has shifted over our heads, exposing its edge on the side the sun is sinking. It's giving a little more light but disastrously causing a massive difference in intensity from the horizon to the centre of the shot for our video. On our cameras, the light completely blows out the foreground – the tracks – which is not good for our video.

17:30. The *Flying Scotsman* is no more than twenty minutes away. The horizon has become incredibly bright as the sun is blazing on the fringes of the clouds, still tucked away and barred from us by the gloom overhead. I turn and face Ryan. 'At least we'll see it now,' I say, trying to bring a bit of positivity to the moment as I can sense Ryan is still a bit down about the day. The cold white light has warmed slightly to a magnolia with the angle of the sun decreasing with respect to the horizontal, starting to set and dipping below the gap in the cloud sheet. The lowest cord of the sun allows gentle yellow light to expose the texture of the clouds: smooth, laminar, almost showing a clear divide between pressure differences in the air. Nearing us, the clouds undulate slightly then tear open above our heads, revealing darker, bluer pockets of cloud that are not touched by the rays. The sun continues its descent. The magnolia becomes clear yellow, blushes a little into pink, and in a matter of minutes breathes warmth on the underside of the clouds in pure shades of orange and red, turning the dark blue above our heads into purple. Ryan and I stand with our mouths agape. The light that was once absent is beaming from the horizon, reflecting off the clouds, washing the valley in a perfect haze.

The *Flying Scotsman* approaches the beginning of the climb. I needn't check Traksy, I can hear it puffing from 2.5 kilometres away. The valley carries the burst of pressure from the boiler, reverberating off the edges, minimizing in intensity but carrying the message of approach nonetheless. Ryan looks at me with his mouth in a circle, breathing repeatedly. The man behind me is crunching around in the bushes – this isn't a shot to mess up and it sounds like he's double- and triple-checking that everything is lined up and set. The two children are hushed, their grandad encouraging them to listen to the locomotive. 'Is it coming, Grampy?'

The exhalations from the *Flying Scotsman* are becoming louder and louder; the distinct firing order of the A3, the class of the *Flying Scotsman*, with its three cylinders, becomes more apparent, cantering. It's very similar to a horse laying down its power on an incline in the way it sounds: 1, 2, 3 ... 1, 2, 3 ... 1, 2, 3 ... 1, 2, 3 ... There are small intervals in the cycle where there is a gap in the noise emitted, followed by a flurry of the three-burst sound, just the same as there is a moment when a horse's legs do not touch the ground when it's cantering. The steam roly-polies over the tree line of the first corner before the locomotive pops into view. A fantastic contrast of cold fluffy white against the warm red of the sunset, emblazoned across the cloud sheet above the train, is captured perfectly on our cameras. I glance down at my screen and it looks fantastic. High-intensity headlights shine powerfully from the front of the locomotive, not quite as majestic as the original oil lamps but a whole lot safer. Speaking of which, the locomotive lets out a little toot in response to the whistle board for the foot crossing, which sets me and Ryan off. We excitedly gasp at its approach.

The repetitious pressure release becomes more furious as the *Flying Scotsman* approaches. *Phutapup phutapup phutapup phutapup* – each *phutapup* happens four times every second, a complete rotation, a complete cycle, pushing, pulling, sucking, blowing; each component experiencing unimaginable amounts of stress and strain through the metal, kept in smooth motion by oils, greases and bearings. The locomotive has passed the foot crossing now, and is nearly at the top of the hill. The side profile is now in view, pistons and rods moving in repetition like a crawling spider, yet confined to the vertical plane, parallel to the tracks. Goosebumps start to pepper my arms, which is often my body's response when I hear a sound I love. Both Ryan and I give the crew a wave. I can just about make them out in their sooty navy boilersuits. A whistle rips out from the top of the locomotive, near to the driver's cab, steam rippling from the outlets. I laugh, gasp and jump all at the same time. I'm completely overwhelmed.

Phutapup phutapup phutaPUP PHUTAPUP – the energy from the *Scotsman* reaches maximum intensity for us, the observers. Eyes and mouths wide open, steam billows from the loco straight up towards us, covering us in a thick fog. A sudden change in the sound from the locomotive as it passes into the tunnel beneath our feet. Dissipating steam allows for the low-frequency wavelengths of the red sunset to peep through as the carriages draw in underneath us. The wisps of steam suck themselves into nothingness, the sound of the whooshing coaches now overpowers the disappearing *Flying Scotsman*, and Ryan and I look at each other in amazement. We wait until the last coach has disappeared, then we both stop recording.

'OH MY GOD, THAT WAS EPIC!'

Chapter Eight ➔

The
Railtour

Location: M4 motorway, 11 March 2022, 16:32

Heavy traffic has built up on the M4 just before junction 18. I'm running late to see Ryan and this really isn't ideal. Google Maps says this jam will add an extra eleven minutes to the journey, with the line ahead going from amber into red, from bad to worse. Before this crept up on me I was only going to make it to Yate Station with two minutes to spare before an empty coaching stock move passes through, relocating for a class 37 railtour tomorrow. Now I'm almost certainly going to miss it. I feel bad as I know Ryan has had a tough time at work recently and I've come a day early to spend some time with him and hopefully help him to feel better. Yate was supposed to cheer him up.

Two class 37s, 37218 and 37425, will be double-heading on the railtour tomorrow, meaning two locomotives are joined together at the front of the train, from Bristol Temple Meads to Bolton. This would normally be an exciting move just to watch, but this time we will be riding on it too. For the first time ever I will get to ride behind a class 37, a growler, nickname coined simply from the sound it makes, the epitome

185

of thrash. The 37s are both owned by Direct Rail Services: 37218 is in the company's typical livery, 37425 is special as it's in the Regional Railways livery, a nod to a past train operating company that used to operate class 37s. The tour has been organized by Pathfinder Railtours. Bashers will be taking most of the seats on the tour, adding to their mileage behind class 37s.

Bashing is a sub-community of train enthusiasm, typically associated with the class 37, and has been going on for decades. It involves purely riding on the train. Not a journey from A to B, not just appreciating the view out of the window, it's the locomotive up at the front that's the core of bashing; leaning out of the window and listening to the pure thrash, typically in the front coach, sometimes two or three people hanging out of the window of the front vestibule. Standing by the trackside or on a footbridge listening to the approach of the locomotive thrashing, the climax and the departure in the distance is a momentary snippet of what it's like to be in the carriages themselves, riding behind the locomotive.

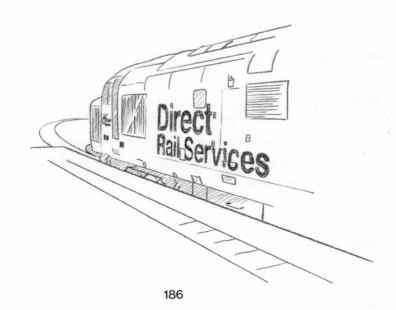

Bashing is different from the trackside experience and delivers the pure intensity of the engine noise, which is only the peak moment from an outside perspective. Of course appreciating the thrash *used* to be expressed by hanging out of the window and flailing or bellowing – in other words moving your arms up and down, sometimes in a chopping motion, sometimes just an arm pointed out straight. Nowadays, the safety of the railway has improved and hanging out of the window is strictly prohibited owing to the risk it poses to life. Under no circumstances is hanging out of the window ever acceptable. Thrash can still be appreciated though. I identify with the bashing community as I love the thrash of a locomotive and I ride trains just for fun. I'd call myself a basher. Others would not. I haven't racked up enough miles behind a class 37; I haven't racked up any in fact. I'm hoping to speak to some well-established bashers on the tour tomorrow and see what they think.

Ryan sounded very upset on the phone. After a petrol stop, I gave him a quick call before I set back off on the motorway. There were a lot of pauses and sighs in response to my apologies for being late. I'm almost certainly going to miss the empty coaching stock (ECS) move to Bristol Kingsland Road, essentially the locomotives and carriages coming down and setting up near the station so that they're ready to go early tomorrow morning, but he is there.

> **Please get a good video so I can see what it's like, mate.**

I ask.

Yep, sure.

Also, remember the XC HSTs cross over at Yate at around 5-ish – we should still be able to see them!

Yeah, sounds good, see you in a bit.

Location: Yate Station, 11 March 2022, 17:18

Huge potholes litter the entrance to Yate Station. I park facing the platform. Ryan is nowhere to be seen.

Hello, where are you, mate, I can't see you.

That was bloody dreadful, no tones and wasn't even thrashing.

Oh dear, sorry, mate. Where are you?

Ryan's supposedly on the platform but he definitely isn't here. I walk down. It's dead, no one.

Ryan chirps up:

Oh I see you, hang on.

He ends the dialogue.

I'm totally puzzled, looking around. Maybe he's in the shelter. Nope. Rotating 180 degrees, I realize the roadbridge that marks the end of platform 1 also marks the beginning of platform 2, which is through on the other side. I see him waving through the mouth of the arch. He really doesn't seem happy – I can tell by the way he is walking. Stairs lead up to the road that cuts through the two platforms. Ryan has already crossed by the time I'm up on his level.

'You all right, Ryan?'

'No, I'm fed up of these miserable drivers not doing tones.' He looks very upset, almost on the verge of crying. Understandably: he's had a tough few days and I know exactly what it's like to look forward to a trainspotting outing all week only for it to fall to pieces when things don't go to plan, the emotion of the week crashing back down again. I briefly put my hand on his shoulder for comfort.

I feel it's best that we take a moment in the car where it's a bit more quiet. We sit there together for a while, talking about the lack of tones, how his week has been at work and how he hasn't had the best of luck recently with trainspotting.

Headcodes

Headcodes, or train reporting codes, allow services to be identified. They are comprised of a single-digit number, followed by a letter, followed by a two-digit number. Each of these components of the headcode gives information about the service.

Classification

1 Express passenger train

2 Ordinary passenger train

3 Parcels train, railhead treatment train, specially authorized freight train or empty coaching stock

4 Freight train with maximum speed of 75mph

5 Empty coaching stock

6 Freight train with maximum speed of 60mph

7 Freight train with maximum speed of 45mph

8 Freight train with maximum speed of 35mph

9 Class 373 or specially authorized passenger trains

0 Light locomotive

Destination Letter

E Eastern	For travel within regions, the letters can be used to identify a destination within the region or a particular route in the region.
L Anglia	
M Midland	
O Southern	
S Scotland	
V Western	

Individual Identifier

As there may be many services following the same route in one day, the two digits after the destination letter are used to separate individual services.

'I don't even wanna go tomorrow now.'

'Oh Ryan, it's going to be fantastic, don't give up, mate.' I remember to check when the XC HSTs are coming. Sure enough they are due through here two minutes apart with the first expected in fifteen minutes' time. 'Look, Ryan, luck's on our side now!' Usually 1V50 from Edinburgh doesn't stick to time due to it being such a long journey; 1S51 from Plymouth is usually more reliable and lives up to the expectation today. Ryan perks up a bit. 'Come on, let's go on the platform and get ready, mate.'

We both get out of the car and make our way over. The sky is clear above our heads but over on the horizon, heavy groups of cumulus clouds smother the sunshine. It's an odd light, not very intense, blue and quite depressing. The end of platform 1 gives the best view in both directions, maximizing the possibility of an HST crossover on camera. 1S51 has picked up a lot of time through Bristol Parkway and looks like it will be here first.

The steady MTU vibration starts to rise from the horizon. I can tell that it's on top notch as the intensity of the rumble sounds exactly the same as it was for me standing on the overbridge in Normanton on Soar, just outside Loughborough. The rain-dense clouds on the horizon have turned a textured baby blue, and wisps of cirrus clouds have caught the sun at a higher altitude to the cumulus puff, turning a fresh yellow. The sunlight has been choked and doesn't look ideal on video. The headlights of 43208 catch the insides of the railheads before it reaches the apex of the corner.

'She's coming!' I shout to Ryan, only a few steps ahead of me and slightly to the left. The lights power around the corner, faster than I expected.

'Wait till it goes under the bridge,' Ryan says with a smirk. Usually the immediate reflection of the exhaust under a bridge or in a tunnel creates a construction of intense thrash. 'That's loud, innit. Is this uphill?'

I nod. 'Yeah!'

The recent rain has dampened the smooth concrete sleepers, turning them into mirrors, the headlights catching them ten or so sleepers ahead of the train. The tops of the railheads offer a reflection too but about two metres behind the reflection of the sleepers as they sit higher off the ground and catch the angle of the headlights at a shorter distance. The snake of the trailing seven coaches and rear powercar have fully emerged from the obscuring corner, following the path the lead powercar forged a few seconds earlier.

'Come on, you beauty!' I shout to the driver, or rather the locomotive.

'Thrash it!' Ryan adds.

43208 passes under the bridge at 70mph, momentarily projecting a superimposed boom of thrash, out the other side. I take a deep breath and exhale, the sensation of the rushing HST reaching my outermost capillaries.

'Nope,' Ryan remarks.

He's sulking at the lack of tones. Tones are the cherry on top of thrash, but the absence of that cherry doesn't bring me down too much, unless the locomotive is coasting, then tones can be the saviour of an otherwise dull moment. It's different for Ryan. I can tell he takes it more personally, as though the driver has chosen not to tone for him specifically, rather than not wanting to do them for enthusiasts in general. In most cases, I think it's more to do with the driver not seeing the enthusiast in time.

BOP ... BIP.

The low tone of the powercar sits softly, the succeeding high tone barks like a snapping Alsatian, making me jump – the extra adrenalin kick, the cherry. It's a bit late now but a thrill all the same, and proof that the driver just hadn't seen us in time.

Coaches rush by. Internally, smoothly rocking through south Gloucestershire; externally, whooshing and clattering past my eardrums, the 23-metre carriages passing by once every three-quarters of a second. 43239 brings up the rear. Chances of a crossover happening now are quite low. I checked just before 1S51 passed through and 1V50 was three signals away. Regardless, I hoped to see the high-intensity LEDs of the lead powercar on the opposite track, approaching as the rear of 1S51 receded from us – the ultimate shot that I am yet to catch. Last year, during the summer when the sun was scorching, Ryan was set up at Blackbridge, Standish Junction, where he caught both CrossCountry HSTs in one shot – the holy grail. At the time there were five HST sets running every day, predominantly on the Plymouth to Edinburgh/Leeds service and return. Now there's fewer, sometimes only two sets running a day, occasionally none. CrossCountry favour the newer Voyagers over their HSTs due to the availability of the Voyagers being far superior; in other words, the HSTs spend more time in the maintenance depots being repaired and are more likely to break down while out on service. Also, an interesting fact I learned recently is that in order to keep the HSTs running in optimum conditions, the MTU engines are required to be run for eighteen out of twenty-four hours a day, I believe to regulate lubrication in the engine. Not ideal when sets aren't on their usual trips back and forth from the depths of Cornwall to the distant cities of Scotland, potentially expediting their decline. CrossCountry are predicted to get rid of their HST sets at the end of 2022/23. It will be hugely sad when they are retired and I must get a crossover shot before they go.

43301 banks around the corner, only two minutes after 43239 disappeared out of view at the rear of 1S51. It doesn't seem to be thrashing: the heat haze isn't flurrying out of the exhaust as it would on full chat. Still at line speed of 70mph, it's approaching us on a slight downhill which probably explains why the driver has drawn back on the power. No thrash then. I can just hear the sound of the rails whirring, wanting to squirm out from the grip of the sleepers under the centrifugal load of the train being forced around the corner through contact of the wheels on the rails. The sleepers themselves are held in position by the surrounding ballast, which is a funny thought when you consider the amount of ballast in comparison to the hundreds of tons of metal flying

---

Britain's Most Northerly Station

Name Thurso

Platforms 1

Trains per day 8

Latitude 59° North

Opened 28 July 1874

Further information Thurso lies as far north as the Alaskan state capital of Juneau and the city of Stavanger in Norway.

above it. Coasting through the platform, the driver lets off a succession of tones: *DU DI DI-DI DU DI!*

'ILKLEY BAHT 'AT, Ryan!' I shout over the noise.

The driver, with a long silver beard and glasses, then sticks one arm up in the air, flailing, the typical expression of a basher, and barrages through the platform. I double over in a laughing fit at the funny tones and the expression of train-spotter appreciation. 'Did you see him, Ryan, he was flailing in the cab!' The coaches rush by as we stand behind the yellow line, blowing my District Line moquette scarf around in the air. It's wrapped around my neck so it shouldn't fly off but I grip on to it tightly anyway. The rear powercar rushes by, 43207 – still no thrash, but the tones made it all worth it.

Ryan's laughing too. 'He did an Ilkley Baht 'at! I didn't see him though, what did he do in the cab?'

'He flailed – you know, what bashers do.' I rewind my video and show him. I got a perfect frame of it. The driver with his arm up in the air, headlights wide open, long silver beard.

Location: Matson, 12 March 2022, 03:50

My alarm screeches in my ear at 03:50. I stayed in a hotel last night in Gloucester next to a dry ski slope only five minutes from Ryan's house. Last night we stayed up to see the last GWR Castle Class HST set at Gloucester Station, 43041 *St Catherine's Castle* and 43162. The fact GWR are naming their Castle Class HSTs after castles and giving them nameplates makes me so happy, making sure the last years of their life are celebrated and enjoyed on their journeys into Wales, through Bristol and down into Cornwall. The Castle Class pays homage to the original 4073 'Castle' class locomotives, built between 1923 and 1950, a majority of them named after castles along the routes that they served. Anyway, it was well worth it last night, especially to see my favourite again, 43005 *St Michael's Mount*. A lot more smart-looking with its large metal GWR logo on the side instead of the mediocre vinyl that's applied to all but three other locomotives. It also sounds a bit more thrashy than the others. Anyway, our railtour leaves Bristol Temple Meads at 06:15. I have to get ready, leave the hotel, pick up Ryan, drive down the M5 southbound, park at Bristol Temple Meads and get to the

Railtours Explained

A railtour is a chance for railway enthusiasts to ride behind classic locomotives that would have run on the mainlines twenty, thirty, forty years ago, and to experience first-hand what would otherwise be memories of the past or Super 8 videos on YouTube. They are an opportunity to see mark 1 carriages from the mid-fifties, class 20s, class 37s and class 50s thrashing away, barrelling along just as they would have done in their prime.

Railtours start early in the morning: departing from London, Birmingham, Crewe, Perth, and heading to locations like Paignton, Llandudno, Carlisle and Shrewsbury, then returning to the original station. Along the way, the routes traverse freight lines, loops and stop in sidings that normal passenger trains would never have cause to – all to please railway enthusiasts. It's a coming together of people of all ages and stripes, to appreciate our country's great railways and locomotives.

platform. I have left a one-hour contingency, just to be completely sure we will be able to get on the railtour.

At 04:10 I turn left onto Ryan's road, mindful of the pothole. My headlights pan across until we straighten out, illuminating Ryan at the side of the road. He's wearing his 'Life is Heaven on a Class 37' t-shirt. I smile: someone's ready for today, I think to myself. It's not my favourite t-shirt of his though, his 'Eat, Sleep, HSTs, Repeat' t-shirt takes the gold medal in my opinion. Today, I'm wearing my pinstripe Aquascutum two-piece, but, annoyingly, last night I discovered two moth holes in the trousers; my British Rail driver's shirt from the 1990s; District Line moquette scarf and some loafers. Ryan hops in. 'Mornin', I didn't sleep one bit,' he says with a sigh.

'Oh dear, I think I got about four or five hours in,' I reply, realizing that I sound a little bit boastful. 'Feel free to sleep in the car if you want now, mate?'

'No, no, it's all right, I'll keep you company.'

It's 05:12 and we arrive at Bristol Temple Meads, more than an hour early, creeping through the car park entrance. Narrow and peculiar, it almost feels like it should have a portcullis. Ticket on the dashboard and our bags collected from the boot, we are ready to start our class 37 bash. For once, I am going to make the most of my full-day parking ticket. We are due to return at 23:10 this evening, using three-quarters of my designated time, instead of the usual one-twelfth or something ridiculous like that.

A GWR guard just starting his shift walks along with us to the platform. We explain the railtour and the significance of the class 37s, unfortunately he won't be able to see it, as his first journey of today is at 05:45. While walking into the main concourse, I suddenly realize we don't have standard, scannable or paper machine-tickets for the gate-line. An interesting situation, as I wonder if the gateline staff will know about the special service or if they'll think we are rowdy clubbers trying to get the first train of the morning back to our sleepy Somerset town. I think the train-related clothing is a good giveaway, particularly Ryan's t-shirt. The gates are open. A shame really: I wanted to see if they knew about the class 37s.

We say goodbye to the GWR guard, who strolls ahead to the underpass. Now we have a bit of free time to check out what's around the station. I love Bristol Temple Meads. Originally a terminus designed by Isambard Kingdom Brunel and completed in 1840, it was augmented through the years to the point where, in 1965, the original terminus ceased to be used. In fact, I just drove across the interior of the old station terminus to get to more parking spaces on the other side, which also used to be a goods yard.

The impressive canopy we know as Bristol Temple Meads now was created as a through line by the Bristol and Exeter Railway, later growing in 1870. In a modern impression of Brunel's metalwork, the arches of the canopy have been obscured by lines of interlinking arched scaffolding while repairs to the roof are carried out. Workers have to traverse the scaffolding above the tracks, so white fire-retardant sheeting provides protection in case any HSTs decide to catch fire whilst below the scaffolding.

Ryan and I walk down platform 4 as we've spotted a special class 150, 150238. It's in the old First Great Western pink and purple livery, the first time I've seen one since 2016 in Frome Station. We move down past the end of the canopy to watch it depart, no tones as it's so early in the morning. It's still pitch black.

06:14, the empty coaching stock is reversing into platform 15. It left the sidings fifteen minutes late and there's no chance the consist is going to be able to stop, get everyone on and depart in one minute. The sky is icy blue in the east, where the coaches are reversing from, with tinges of peach on the horizon. The light catches the sides of the carriages as they march around the tight radius. Silence as they draw in; it's very unusual not to hear any kind of exhaust burbling, almost like a silent film. None of the interior lights are on, the only sign of life is the flashing red tail light.

37425, *Concrete Bob*, makes an appearance first, attached to the first coach in the train, its smart yellow snow plough and blue and white regional railways livery looking fantastic, so clean too. The nameplate of 425 pays homage to Sir Robert McAlpine, a pioneer in the concrete industry. On the other side of the locomotive, there is another name plate with Sir Robert's full name, which I think is a lovely gesture. 37218 follows 37425 reversing into the platform. Not the prettiest of locos, especially with its low-hanging BMAC light modification. It's apparently knackered as well, and this will be its last trip, but I'm fairly certain that's just a rumour. Both of the locomotives are dormant, idling as they settle into the platform, brakes squealing as they come to a stop.

I'm keeping an ear out for any distinctive difference in the exhaust note of the 37s, trying to suss out any issue with 218. They both growl away. Nothing unusual. 425 is more aggressive whilst idling and coughs out a little more clag; it's certainly the more desirable locomotive. More than fifty fellow train enthusiasts arc around the locomotives to get photos, unfortunately for us 37218 is too far forward to get a shot of the front and 37425 is sandwiched between 218 and the coaches. 'Ah well,' I say to Ryan, 'let's get a seat and settle down'.

We will be riding along in mark 1 carriages today. Built from 1951, they were made in response to nationalization after the Second World War, fit to be compatible with existing stock around the country; stronger, safer and more ergonomic. The windows have cream, tethered curtains attached to a rail between the top of the window and the grated luggage rack. Wide and tall, they are situated perfectly between seats to offer passengers what they want: a nice view, instead of being packed between ironing boards where the only thing to look at is a window divider – class 800, I'm looking at you.

Varnished mahogany wood borders the carriage; dark and streaky, it reflects the dim morning light well. The seats are like armchairs but for two people: blue with a William Morris style pattern of golden maple leaves. A thick divider, upholstered in the same material, sits in between the two-abreast armchair, ideal for a bash snooze. The seats alternate in orientation, lending themselves to a sociable journey, with a vinyl-topped table between the facing seats. Mahogany-framed glass panes divide the back-to-back seats.

Ryan and I sit either side of our table, upon which there are booklets, carefully put together, detailing our journey ahead. I flick through the pages: beautiful photos of previous class 37 railtours, some taken by Jack Boskett, a well-known railway photographer.

My eyes are particularly drawn to the gradient profile page; it shows a 2D cross section of the gradients we will run up and down on our journey to Bolton. Lines divide each gradient change, causing clusters of ratios at certain points. Around Gloucester Yard Junction for example, the different gradients have been reduced to intense lines bordered by 1:-417 on the Bristol side and 1:344 on the other side.

'Look, Ryan, at Yate, remember we thought 1S51 was thrashing because it was going uphill and 1V50 was cruising downhill?' I point to Yate on the diagram. 'It's actually the opposite to what we thought.' Ryan doesn't seem to be as interested. I'm totally fascinated by this, it's something I could look at for hours, just like maps. I think it satisfies the same part of my brain. Scanning across the route, I realize that tunnels are also noted, with black blocks above the line, lines on either end, connecting to the gradient line, marking

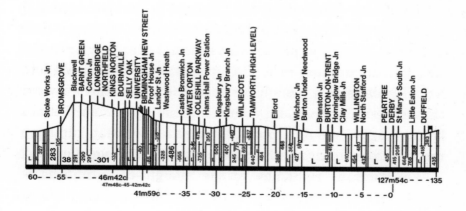

the beginning and end of the tunnel. 'Lickey Incline, Lickey Incline, where's that?' I mutter to myself. 'Just after Bromsgrove, ahh, gotcha.' It's the steepest part of our journey, the steepest incline on the mainline network, in fact: 1:38 or 2.65 per cent. For a car or lorry, that's not an issue and can be dealt with by dropping a gear. For the train, however, it can be particularly tricky with a reduced amount of driven wheels, especially a freight train.

The issue is overcoming the influence of gravity. Usually the locomotive, with zero gradient and at low speed, only has to contend with contact friction, which for steel on steel over an area similar to a 20p coin per wheel is very little, making it a great way to transport heavy loads. When a locomotive has to overcome gravity's opposing effect on its motion, low contact friction makes it hard to create traction via such a small contact area.

For the Lickey Incline, the answer to the slow or failed ascents is the use of 'banker' locomotives. Attached to the back, they assist with the incline by providing extra grunt. In the new age of diesel, this used to be the job of a pair of class 37s working in conjunction with their hydraulic counterparts, class 35 'Hymeks'.

The funny thing about the Hymeks is that the terminal speed of a typical train up the Lickey Incline, with a Hymek assisting, would be just in the spot between the first and second transmission ratio. So, at maximum power, the loco would indecisively switch between the two ratios. 'Owww, this is getting tricky, I'll switch to this one … ahh, that's better. Oh, we should probably change up, we're in high rpm range. Oh, actually that's quite tricky, lets change down, oh …' I'm laughing right now at the thought of an indecisive Hymek getting grumpy at the incline. However, they solved this by locking off the lowest gear ratio, which is ironically the best for tractive effort, meaning the Hymeks dedicated to 'banking' had to slog along in the second gear. 'Banking' locomotives are still used to this day on heavy freight trains up Lickey, operated by DB Cargo. 37218 and 37425 shouldn't have an issue though.

Our window is open. One-third of the way up there's a metal divider in the window that houses the sliding mechanism. In the closed position the top third of the window is divided into three equal-sized sections, with the centre section split in two down the middle. This centre section is clasped together by a sprung latch, heavy and slightly tarnished — many hands have applied force to it, much like the twistable door handle. To open, the latch must be lifted and, on the side that the latch pivots, pulled across. This opens one half of the centre third to the cool breeze of the morning. The other half can easily be pushed all the way across. In less safety-conscious times, this aperture would have been the perfect size for a basher to protrude a head and an arm. Certainly not nowadays though as it's strictly prohibited. Proportions of the mark 1 windows are perfect. Nothing has come close since, apart from those of the mark 3 carriages, but I feel that's an unfair comparison and an altogether different category.

Rather luckily, we are on the right-hand side of the train so, due to the curvature of the track, we can see the loco-motives at the front through our window. Positioned three carriages from the front, we can really appreciate the thrash.

37218 and 37425 open up their type 3 English Electric engines. I immediately gasp. Ryan opens his mouth wide. The stillness of the morning is ripped to shreds by the two brutes, 425 more brutish than 218, lots of clag and absolute hellfire powering out of the exhaust pipes. I can't help but laugh under my breath, goosebumps all over my body. The thrash builds up as the locomotives gain traction, accelerating out from the platform a lot faster than I expected. I'm astounded by the echoes that I can hear bouncing off the surrounding industrial buildings, they're so far away but so raw. I'm standing up, my ears are closer to the window, but I dare not get any closer. Eyes closed and my brain focusing on one thing. I sit down again on a beanbag of thrash; comfortable, even though my legs are not fully outstretched because my knee sits under the top of the table and the seat is too far forward on my calf. 06:45 now, we arrive in Bolton at 15:30, I couldn't be happier about it!

We pause at Bromsgrove, enough time to get out for some photos of the locos before barraging up Lickey Incline. Ascending the footbridge slightly, I get a fantastic shot of both locomotives. 218 is a little too far forward, but I get both locos with a decent amount of carriages hooked on behind. Jack Boskett, the photographer, also offers to take a photo of me and Ryan holding the Pathfinders nameplate in front of 425. Very satisfied, we return to our carriage.

The initial onset of thrash is tentative, to maintain traction. Second and third blips extend for a little longer and are deeper. Fourth, the distinctive turbo whistle builds up, powering along with rumbles. Fifth, the thrash remains constant from here, building and building. 'This is it now!'

I'm pressing my face against the glass and looking up along the track to see the locomotives. The radius leaving Bromsgrove is a little too large to see them completely, but the refractions within the double glazing of the windows convey blurred, distorted colours of the class 37s, particularly 425 in its distinctive blues. They rise suddenly after passing under a bridge. 'Up we go, this is it!' The first, second, third, fourth coaches follow the 37s, looking ahead to the massive effort required from the beasts. Just as I turn to look at Ryan, a train blasts by in the opposite direction.

'HST!' Ryan beat me to it. He saw it, I heard it. A brief MTU burst through the open window. The rear power car follows, leaving silence just for a moment. My ears then adjust to the now reduced volume and I can only hear one thing now, an English Electric anthem for the rest of our journey.

Chapter Nine ➜

Bashing

Location: Bolton Station, 12 March 2022, 15:32

We've arrived into Bolton slightly later than booked. My friend Luke, who became a trainspotting partner during our time together at university, has come to meet us. He's driven from Stockport, not too far, and has parked by the Sainsbury's next to the station according to his text message – quite convenient as I'm starving and need a sandwich.

Ryan and I run across the pelican crossing just as the walking people indicator turns from green to red. 'Peugeot 207 in silver, mate, keep an eye out for it,' I say to him. But I needn't look any further as I can see my friend standing by the entrance. 'There he is, Ryan.' We walk over and I introduce them to each other, quickly dashing inside without being rude to get a meal deal. The tour departs in fifteen minutes so we need to be quick.

Running down the stairs to the platform I can hear the locos powering up. They must be running around to reattach to the front. Sure enough, as I'm five steps from the bottom, 37425 finally leads 37218 without any carriages, shackles free, maximum acceleration, down the adjacent platform to the one we arrived in on, which still hosts the stationary carriages. They run along to the first set of points a few hundred metres down the line to then switch back on themselves to reattach to the train at the other end. We were closer to the front on the way in, so our seats will be towards the back on the way out.

Platform 5 now carries a concentrated mass of spotters surrounding the locomotives. There looks to be plenty of room on the platform in front of 425 so I should be able to get the photo I've been looking for. Luke, Ryan and I try to find a spot in the crowd. Not wanting to get in people's way, I crouch down between their legs and snap a photo of the two heritage locomotives: the orientation of the lights has switched from trailing end: red to leading end: high-intensity white, the headplate has switched ends, perfect shot. Coupling up the locomotives to the carriages divides the group of spotters as some of us split off to see it. A particularly favourite moment for me is when the locomotives shunt on to the buffers, compressing them and sometimes compressing the slack in between the coaches. DRS workers set to work on attaching the brake hoses and couplers. Luke watches next to me.

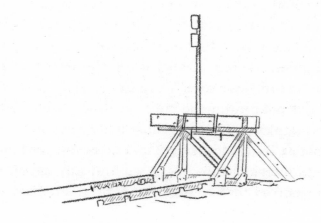

My trainspotting friend James approaches from behind. I first met him on the journey of the last HST departing from London, he's a nice chap and loves class 37s as well. He mentions he's in the carriage right behind the 37s for the journey back, the prime bashing location. We chat a little bit about the journey. We booked separately and are sitting apart, so he kindly invites me to his carriage, the one right behind 37218. I politely accept his offer, saying we will come up a little later in the journey. Ryan and I head back towards our coach, ready for departure.

Luke stands on the other side of the window to watch the departure. I give him a wave. My meal deal chicken and stuffing sandwich having already been consumed, I'm munching on some Quavers as we slowly pull out of the station. It's considerably quieter in our carriage now that we are further away from the locomotives. The return leg will be shorter as we won't be running as far around the Pennines, which on the outbound journey, once we touched Manchester, took us up to Huddersfield and through the valleys, back to Manchester Victoria and out to Bolton. On this leg we will be heading down to Derby as soon as we run through Manchester.

The first hour of the journey provides an opportunity to relax and talk with Ryan about the day so far – the locomotives, the scenery, the sidings and the trains we have seen along the way. A highlight was, while passing through Birmingham, seeing the new car train run over the top of us around the Aston area, just after the big intermodal Freightliner depot heading north. I mulled for a while over the route it was taking. I thought they headed from the new Toyota depot at Toton, down the Midland Mainline. Perhaps it went on a diversion. Cars made in Toyota's factory in Derby are transported to Toton, 18 miles by road. The cars are then loaded on to the double-decker train that runs from Toton depot down to Dollands Moor, just outside Folkestone, where it then transfers on to the Channel Tunnel through to Valenciennes, in northern France. Toton depot was a place I used to go to relax while at university. It was a forty-minute cycle away but the views were well worth it, with lines of stored class 60s in EWS livery – I think there's around twenty-four of them – sent there after the introduction of class 66s, which EWS favoured over the class 60s, leading to almost seventy-five per cent of the class 60 fleet going into storage. Over by the traction maintenance shed there would often be a gathering of class 67s, DVTs and class 66s. The longest freight train on the network, the High Output Ballast Cleaner (HOBC), was a frequent visitor too, hauled by two Freightliner class 66s on either end. A large hill runs up the east side of the

depot providing perfect elevation to look down on the workings of it from above. Lines of young birch trees dominate the foreground of the view. Here, the sidings used to extend across at least twelve more road (tens of tracks) widths, when shunting goods was a lot more fundamental to freight distribution on the UK networks. Previously, freight trains would comprise wagons with goods from varying origins and companies, both privately owned and state-owned. Milk tanks, food stuffs, mail, it would all be linked together on the same train and would then have to be broken down in dedicated marshalling yards where the trains would have to be shunted back and forth, organizing and arranging for the freight to go on to its next destination. These marshalling yards were created across the country, tens of tracks wide and sometimes miles long.

At Tinsley, Sheffield, a forward-thinking system was devised in the 1960s to make the sorting of the freight easier. The composite freight would arrive just before the top of the 'hump' at the beginning of the yard, and shunters would run through the consist, noting down where each wagon was meant to go and how many of the same freight were directly coupled up with it, making decisions on where to uncouple the freight based on this. Uncoupling complete, shunters would then return to an office where they would translate the data of the train into code, with each destination and type of freight having its own combination of numbers, the amount of wagons noted and those requiring particular attention owing to weight of contents marked with particular numbers too. This would be transmitted to the white control tower, where the 'hump inspector' would receive the information on punched tape, giving them a simplified representation of what they could see before the 'hump' from their elevated viewpoint. Once completed, the tape was fed into a point operating machine that read the punches and based point operation on that. The hump was very literally a hump. Shunting locomotives would push the wagons from the rear at a slower than walking speed, up to the hump, where all the roads at the beginning of the yard gradually converged. As the first wagon (or set of wagons with the same code) reached the crest of the hump, gravity's influence carried the wagon(s) away from the rest of the train and down the other side of the hump to its set road. Passing the points, they were then set for the next wagon automatically. The speed of the wagons was controlled by pneumatic pistons with rounded contact surfaces that sat on the interior of the track. Contact would be made between this and the interior flange of the train wheel. Too fast and resistance would be provided by pressure from the surface on the approaching side of the wheel, working against the inertia. Too slow and pressure would be provided on the trailing side of the wheel, assisting the inertia of the wheel.

Unfortunately, this ingenuity fell on the wrong side of demand and the development of freight distribution. Containerization was starting to take hold in the 1960s, growing more through the decades, leading to the development of intermodal freight hubs where containers could be loaded and unloaded directly to and from waiting lorries, removing the need for dividing and sectioning the train. Marshalling yards across the UK fell into disuse and became depots where rolling stock was stored, often only using a fraction of the roads available. Lines of track were torn up. The scars are still present at Toton today. The birch trees have grown from where the tracks once stood, in lines. A three-spike fence separates two states of dereliction: one, on the side that's open to the public, scarcely resembling the industry that used to exist there, the other left as it was to decay – rotten sleepers, rusty track and graffitied postal carriages.

Climate change will either force freight off the road or force freight by road to become less polluting. In either case, I feel there is a new era of freight distribution approaching. It is difficult not to reach the conclusion that rail is the answer. Intermodal freight distribution hubs are halfway there, connecting rail, road, sea and air freight through universal containment, which is greener and more efficient. The distribution of freight by rail removes 1.4 million tonnes of CO_2 each year compared to road transport, and this figure will only increase as more electric freight locomotives are introduced. From there, rail could extend further, as has happened in Germany, where pantographs have been installed on the autobahn. It's only a matter of time before the inefficient high friction of tyre and tarmac is replaced by rail and

Britain's Most Southerly Station

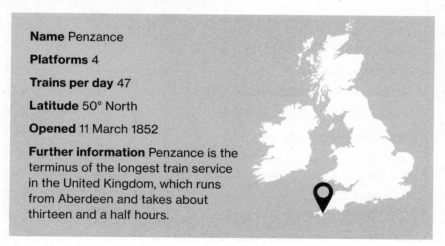

Name Penzance

Platforms 4

Trains per day 47

Latitude 50° North

Opened 11 March 1852

Further information Penzance is the terminus of the longest train service in the United Kingdom, which runs from Aberdeen and takes about thirteen and a half hours.

wheel, steel on steel, at least on one lane of the motorway. So, potentially the ingenuity of the marshalling yard could be called upon again in the future, where track is used to lead from distribution hubs to the destinations of the freight via rail, perhaps to city-based hubs, before only the final miles of a journey are completed by road.

Passing Earles Sidings again near Buxton, I remember that Gordon sometimes drives up here. He does the Earles Sidings cement tanker job, usually in a Freightliner class 70. It's almost impossible to see a Freightliner class 70 around the Whatley and Merehead Quarry area so I imagine he enjoys mixing up the traction a bit. Interestingly, they have a few class 20s on shunting duties at Earles Sidings, which I point out to Ryan as we head past.

Class 70

Build date 2008–2017

Total produced 37

Number in service/preserved 37

Prime mover GE PowerHaul P616

Power output 3,690bhp

Maximum speed 75mph

Current operators Freightliner, Colas Rail

Nicknames Ugly Duckling

We decide to go and see James up in the coach behind the class 37s. The light is fading now and the view out the window is becoming dark blurs. We leave our bags on our seats and make our way to the door. Manually operated sliding doors between the passenger compartment and the vestibules feature in all mark 1 carriages. A vertical metal handle requires pulling anticlockwise to around 45 degrees and then the door is free to be slid open. In the vestibules, the mahogany wood continues with access to a toilet. Here the noise of the train is at its loudest as the interconnecting gantries have comparatively less sound-deadening. The sound of the carriages is almost like a loud, continuous exhale, whirring along with the occasional clatter that travels down the train as it passes over the points. The first clatter arrives from the end of the carriage in front, then the second clatter on the first wheelset of the carriage of the observer hits at an increased volume. The wheelset nearest the observer's position creates the most noise before the sound reduces as the intentional gap in the track interrupts the rest of the wheels behind you. Very much like a singular wave: *clck clck, clCK CLCK ... CLCK CLCK, CLck, CLck ... clck clck, clck clck.*

I open the door to the next carriage along – similar decor and demographic: a few families and some elderly couples out on an excursion. The next carriage along has different seat upholstery and a different layout, more like benches with blue and black hatching. There is a buffet in this carriage selling a selection of crisps, drinks and merchandise. Next to the buffet is a small table with a keg of beer on it, plastic cups stacked high; it looks like the top few cups might topple over at the next set of bumpy points. A piece of paper Blu-Tacked to the edge of the table states that it's £4 a pint and it's some sort of ale. There is a queue of three people waiting their turn, the man at the front impressively looking like he's about to carry three pints of beer at once. There's some grumbling as the man pouring the beer says that he thinks it's about to run out. I'm assuming they have a back-up keg but it's probably in the kitchen or support coach towards the end of the train. The next carriage has blue cross-hatched seats too. More train enthusiasts in this one certainly, the more traditional bunch I'd say. Vinyl tables have route maps open, notepads out, thermoses and cups of tea poured nicely. Knitted jumpers, jackets and flat caps. A few of them are asleep.

We are nearly at the front of the train, the carriages transitioning to more of what I know as the interior of a mark 1 carriage, the same layout as the carriage Ryan and I are seated in but with walls and dividers, vinyl-covered as well as the tables for the sake of durability. The seat moquette is more of a jagged greyish blue. Strip lights run down the centre of the carriage surrounded by opaque plastic, homogenizing the light in the carriage and giving it a creamy tone. This is the territory of the bashers, situated right behind the engine.

My heart starts to beat a little bit faster as I know that train enthusiasts have mixed feelings about me. The videos I make are watched by people who predominantly don't have a passion for trains. From what I've read in the comments on those videos, my average viewer watches them to connect with the sense of joy that I'm experiencing in the moments I share, to see me fulfil my passion and have a fantastic time doing it. Thanks to social media platforms, the videos can be distributed widely if they are enjoyed by the initial sample of people they are released to, the followers of my channels. Trainspotting videos, or train-related content, are very successful on other platforms like YouTube, produced by the likes of Geoff Marshall, Ruairidh MacVeigh and Jago Hazzard. I watch their videos to relax. They are thoroughly entertaining, informative, and rack up high views consistently, but I suspect their audience has a medium to high interest in transport and trains due to the factual nature of the films they create. Thus they are highly regarded and appreciated in the community. With my videos I take a less factual approach, simply documenting the moments I experience, as seen through my lens, whittling down a whole day's worth of trainspotting to the best bits, which are usually moments of pure joy. I realize this sample of my day can then be seen as a representation of the whole time, bounding along laughing in a state of ecstasy, because these are the moments I show in every video. However, there are of course calmer moments, moments of boredom, sadness and frustration, and times when I just want to go home.

From this concentration of my happiest minutes and seconds, viewers from outside the community are left thinking that every trainspotter is like me, and that trainspotting itself is the pursuit of those euphoric, thrilling encounters with trains. But just like the wider world, trainspotting and trainspotters are a totally mixed bag. Everyone has their own priorities and passions, and their own way of enjoying the pastime. So having my impression of trainspotting amplified to the world frustrates some people within the community because, like me, they hold their version of trainspotting close to their heart. Seeing a different form of train enthusiasm inadvertently representing what the whole community is like must be frustrating. I have only come to understand this point of view after reading forums and hearing what Ryan tells me they've been saying in the Facebook groups. On the other hand, I make my videos because I enjoy my trains, and I enjoy making videos and making people smile and laugh. Colliding all those components together has created something that I'd never have imagined – a community of joy. However, for the first time, I am knowingly walking into a carriage where there may be people who don't like me. Part of me wants to head back to our seats at the quieter end of the train, but I am really keen to hear what the thrashing is like in the frontmost vestibule.

The second to front carriage isn't so busy, populated by a similar group of older enthusiasts like before, but I can see through to the front carriage and it's packed. I move forward to grab the handle of the door leading into the prime bashing domain. I slide it open and move in. All of the windows are wide open. Every seat is occupied and there are a few people standing in the gangway. Faces at the nearest two tables look up then look down – these guys are younger, around my age. It goes quiet near the area of the door. I walk through.

'You know you make my life hell at work?'

I turn to my right. My heart starts thumping hard. A basher around my age is sitting in the corner.

'No, no, don't listen to him, he just says what he wants,' a voice pipes up from behind me.

I turn round to see who said that – a chap with blond hair. I turn back to look at the person who made the comment. 'Well, I'm sorry that my videos have caused that to happen to you.'

'Yeah, well, my work colleagues just think I'm like you now because they know I like trains.'

'I'm sorry about that, you know … but I'm just having fun and making videos.'

At this point I decide to move further down the carriage because I don't like the feeling where I am. I examine the next two tables along on either side: the group is a little bit older, a few women too. One of the guys notices me – 'HERE HE IS!' A few others turn round. 'WUHEYYY!' They are drinking beer but seem friendly and genuinely happy to see me. I am quite confused as my heart is still thumping and I feel a little bit like crying but all eight of the people on the two tables are saying hello to me and asking how I'm finding the excursion. This turns my feelings around. They ask for a picture, so I crouch in the middle and the folks around me get closer together to get in the photo. I'm slightly concerned that the flip-phone case camera hole is slightly obscuring the camera because the phone has slipped down a bit but they take the picture anyway and seem to be happy with it.

'That's going straight on Seminars,' one of them says – I assume a bashers Facebook group.

'Thanks, folks!' I say with a smile to the people on the second row of tables, and continue walking down.

The third table doesn't take much notice, neither does the fourth. They all seem older, a few route maps here and there, portable chargers and Domino's pizza boxes. Must've ordered ahead before we got to Bolton; definitely not enough time for them to be ordered and prepared in the time we had at the platform! Most of the people in this carriage are drinking the ale from the keg but there are a few shop-bought cans too. As I negotiate my way around the odd person in the gangway there are a few mutterings, smiles and stares but no one has expressed their dislike, which is a relief.

I can't see my friend James though. The frontmost vesti-bule looks packed and I assume that he's in there. The glass in the door allows me to see into the area and gives me a heads up that opening the door might make people annoyed. A few give me a deterring look, but I suspect it's because it's quite busy. I open the door and purse my lips in a smile, almost as a 'sorry' facial expression. Most of the bashers here are in their forties, some standing close to the windows on either side which are slightly ajar. They are silent, enjoying the thrash.

'Oh, there's James, Ryan.' He's one person away from the window, probably hoping to secure his spot at the front soon, but definitely ensuring not to get too close to the small gap where the window is letting the sound of thrash in. I smile and nod, he nods back with a raise of his hand. There's a tighter group of bashers in the middle engaged in enthusiastic discussion, though I can't tell what they're talking about. I duck through a gap between one of them and the toilet wall. My back is against the wall now, and I'm shuffling. A space opens up. Ryan scoots through too and we find ourselves in between the two toilet doors. Towards the locomotive, the gantry that would lead to another carriage is locked shut and secured. I crouch down and look through the keyhole.

'There she is, Ryan – 37218.'

I put my ear up to the hole but all I get is persistent wind noise buffering against the door. Ryan switches with me and has a look through the hole too. He giggles.

Mild claustrophobia starts to creep in as I realize that we are penned in on all four corners. The gantry door, two toilet doors and a group of fortysomething men confine us to a space not much bigger than a bin lid. We have to stand in a staggered formation to fit. The bashers talk purely about class 37s, sharing pictures of excursions in the Highlands, presumably years ago because they're talking about locomotives that have since been scrapped. They seem to be reliving memories that they share together, aided by photos. All laughing together. I want to join in but of course I know that the conversation is personal and I certainly have no part in it. One of the men looks quite familiar, with a purple shirt, waistcoat and a long beard, but I just can't quite connect the dots in my mind. They haven't acknowledged us, which I'm pleased about because I was worried I'd get a frosty reception.

I start to relax. My attention turns to the sound of the locomotives that were previously coasting when we squirrelled our way into the vestibule. They're burbling along now at their idle rpm. I think we're somewhere near Water Orton in Warwickshire, according to a conversation I heard the tail end of by the window. It feels like we've passed a yellow signal aspect as the brakes are being applied gently, squealing beneath us. I check on Ryan to see if he's OK being in the confined space. He gives a thumbs up with a smile, then mouths 'I'm hungry'. I nod and try to speak over the background noise: 'We can go back to get some food in a minute, let's hear some thrash first.'

A few minutes later, after a brief pause, we are off again.

'THRASH! THRASH!' Ryan shouts.

'Come on!' I join in.

Thrashing builds and builds. I look down and pull up my sleeve: I've got goosebumps, making all the hairs on my arm stand on end. I show Ryan and he laughs. Screaming from the turbocharger picks up and powers along in unison with the growling exhaust. I'm absolutely buzzing. My ears do pick up an unfamiliar sound though – unfamiliar when it comes to class 37s but familiar in isolation otherwise. I think it's the sound of the alternator. It reminds me of an EE507 traction motor sound from Southern class 455s. Some of the bashers near the window comment that 37218 sounds buggered; one of the bashers in the middle group comments that he thinks the alternator is on its way out. Perhaps that's why it sounds unfamiliar!

I'm talking to Ryan about the alternator when one of the bashers turns to him. 'You should like him, he drives Cross-Country HSTs.' He's pointing at the man in the purple shirt and waistcoat, and he taps the man, who turns round.

'Oh my God!' I mutter. I turn to Ryan and he's realized the same thing. Without saying anything to each other we agree on who it is. It's the HST driver from yesterday evening who gave us the Ilkley Baht 'at tones and flailed as he went by. Both Ryan and I try to say the same thing at the same time: 'Were ... weren't you the HST driver yesterday who gave us Ilkley tones through Yate?'

We must've both sounded overly excited because I think he's a bit embarrassed. He is with his mates, and two young, inexperienced bashers are probably cramping his style a little bit. He nods. 'Yeah, that was me. Had a few pals in the carriage behind me coming down for the tour, but I don't usually do tones like that.'

We nod excitedly.

'Especially when kids at stations do this,' he adds, doing the pull-the-lever motion for tones. Universally this gesture is shunned in the community because people say that it seems like it's begging for tones; clearly this driver isn't a fan of it either.

'Lucky we only gave you a wave and a thumbs up,' I joke.

He nods, and with a brief smile turns back to his mates.

'I desperately want a picture with him, Ryan,' I say, lowering my voice a bit. 'Do you mind holding my phone ready for one? I'm just going to wait for the right time.' Ryan takes my phone and I position myself so I can ask him at the right time in the conversation.

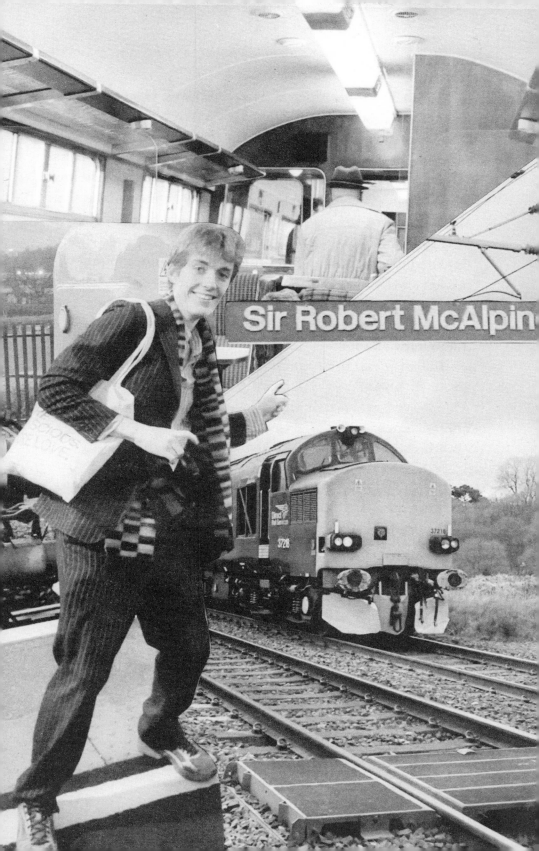

Sir Robert McAlpin

My plans are scuppered when someone comes into the vestibule to use the toilet. I reverse back into the corner with Ryan so he can open the door. He seems a little bit drunk but thanks us for moving out of the way. The door opens, and we have all underestimated the size of the door and how much space it needs to open in order for someone to get in. Mahogany wood is pressed against me as the man tries to fit through. He does so, but now the centre circle of bashers has readjusted, making it quite hard to find a good angle for me to ask the man in the purple shirt for a picture. They're looking at an old image trying to work out who someone is in it, and the conversation drops for a moment. I take a breath – now's my chance.

'I don't mean to interrupt you, but do you mind if I get a photo with you?'

'Yes, you've got one already,' he laughs along with some of his friends.

'I haven't?' I say as light-heartedly as I can.

'You have, when I was in an HST.'

'I'll go away as soon as we've got one.'

'Like for ever? Never come back? On trains, ever?'

'You'll never see me again,' I quip back, sensing the humour in his tone of voice. A few of the guys around him laugh under their breath. The conversation ceases at this point. I can tell that I'm really cramping his style so I step back a bit, accepting defeat.

'Anyway, pinstripe kudos.' He points to his trousers.

'Eyy!' I reply, lifting my leg up and also pointing at my pinstripe trousers. I understand now that this man is actually quite nice!

Ryan and I sit back down at our table after requesting some dinner. I've ordered roast beef and Ryan has gone for chicken curry. It's delivered in a paper box with cutlery. I eat mine as quickly as I can because I want to return to our spot at the front before Birmingham New Street.

Back at the door of the frontmost vestibule, the bashers our age from the first table are in there now. Feeling slightly less trepidation than last time, I slide the door open and notice that the older bashers have retreated to one of their tables. I didn't notice them as I was walking through. The younger bashers groan audibly as Ryan and I walk in.

'All right, guys?' I say timidly.

'All right,' they reply.

After breaking the ice with them, we talk about 37s, bashing, heritage lines and my motives for my videos. It becomes clear that they thought I was someone taking the mick out of trainspotters, parodying them on camera. After understanding that I share the same passion as them, we get on quite well. Coasting into Birmingham near the Freightliner depot before New Street, the blond one turns to his friend: 'Yeah, he's a crank.' Crank is slang in the community for someone who loves locomotives. Hearing him say that means a lot to me because it is the first time I have come across a group that didn't feel warm towards me at first but come to accept me as one of their own.

Location: Birmingham New Street Station, 12 March 2022, 20:38

It's only Ryan, me, James, a few other spotters our age and some older bashers in the front vestibule now. We're nearing the end of a five-minute stop at Birmingham New Street, I think to change drivers. I managed to get another picture of 37425 as well, this time in the dark with the chaotic mirrored reflections of New Street in the background. Climbing back on board, I found a spot next to an older basher who has gelled hair and is wearing a Stone Island jacket. He's leaning with his back against the wall, slightly crouched so he can see out of the window. I'm leaning against the opposite wall standing a little higher than him. We don't make eye contact. An Avanti West Coast Pendolino stands next to us. The passengers inside look through the squat windows, puzzled by the older carriages and heritage locomotives of our train. Some take pictures.

37425 and 37218 growl deeply. I haven't heard them pull away from a standstill yet, at least not from our new forward position. The basher next to me is filming from

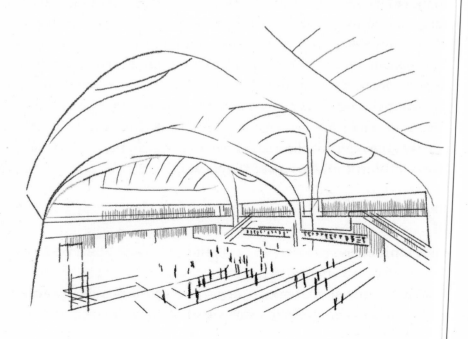

inside the carriage so I'm trying to stay quiet so I don't ruin his video. Clag pours out from the top of the locomotive. I can tell because the reach of the platform lights is eclipsed by the wafts of soot, dulling the light around us and creating heavily contrasted lines between the atmosphere on our platform and the well-lit space beyond. As we pass the front of the Pendolino, I see the driver has his near-side window open. He's watching us depart while nodding appreciatively at the roar that's so uncommon nowadays in the hub of Birmingham.

The engines stop thrashing as we negotiate sets of points leading into the tunnel just after the station. The inside of the adjacent track is bright green, the signal lights reflecting off the oily metal. It's the only thing I can now see, peering through the window. Green suddenly turns to red as the locomotives pass the signal – surely he should be back on it by now. If I remember correctly, the gradient diagram displayed quite a rise after New Street leading up to University Station. 'Come on!' I'm getting a bit overexcited under the suspense of thrash but I quickly reel it in, remembering the man videoing. As we screech around the left-hander, the inside walls of the tunnel start to light up. I suspect headlights from another train are catching the moisture on the inside of the tunnel. There it is! 'Oh YES!' The thrash starts to build again while we're still in the confines of the tunnel. Growling. Growling! 'COME ON!' I internally go into a frenzy as the class 37s gun it out from the mouth of the tunnel. The reverberations inside the enclosed space at top notch make me hyperventilate and everything goes jittery. I'm in seventh heaven. The thrash is released as we emerge from the tunnel, reducing the volume, but it is almost instantly one-upped as we pass the class 323 that was beaming into the tunnel. Unhinged combustion

Railway Enthusiast Terminology

Bash	To travel on a particular train in order to experience a desirable locomotive in action
Basher	Someone who enjoys bashing and has clocked up many miles behind desirable locomotives
Bowled	When the view of a train or locomotive you are hoping to see is blocked by another train passing in front of it
BR	British Rail
BREL	British Rail Engineering Limited
Cab ride	Riding in the driver's cabin at the front of the train
Choppers	Nickname for class 20s, due to the chopping sound of their exhaust
Clag	Soot that is chucked up in the exhaust, creating a dark cloud
Crank	A railway enthusiast
Crompton	Nickname for class 33s, which have Crompton Parkinson traction motors
Deltics	Name for class 55 locomotives, which have Napier Deltic Engines
DMU	Diesel Multiple Unit
Duff	Nickname for class 47 locomotives
ECS	Empty coaching stock, the train is without passengers and is not in service
EMU	Electric Multiple Unit
Flail	Waving an arm up and down in appreciation for a good locomotive

Flying Banana	Nickname for the New Measurement Train (NMT) as the whole train is yellow and can travel at speeds of up to 125mph
Gen	Information on when and where a locomotive is going to be or what service it will be running
Gronk	Nickname for class 08 shunters
Growler	Nickname for class 37 locomotives
GWR	Great Western Railway
Hellfire	A phrase that is usually shouted when an amazing trainspotting moment occurs on the railway
Hoover	Nickname for class 50 locomotives
HST	High speed train. Debatably, the only time to use this is when referring to the class 43
Insects	Young railway enthusiasts
Kettle	Steam locomotive
Light engine	When a locomotive is travelling by itself without a train
Railtour	A private charter train run specifically for railway enthusiasts
Sulzer	Nickname for locomotives with Sulzer engines, such as class 47s
Thrash	The name of the noise produced when a locomotive is powering up, particularly by loud locomotives like class 37s
Tractor	Nickname for class 37 locomotives
Warships	Class 42 and class 43 locomotives
Westerns	Class 52 locomotives

inside the engines reflects off the parallel unit at a distance of only a metre, bursting into our carriage, even louder than the sound in the tunnel, continuing down the full length of the two three-car units. We're picking up speed fast, passing the end of the 323 carriages, past arched retaining walls and back into another tunnel, emerging at Five Ways Station. At this point we are doing around 30mph, both engines still at full chat, as we pass a two-car CrossCountry class 170 sitting at the platform on the other track. There is another bashful reflection of thrash as we pass the unit, although the wind noise is starting to dampen the purity of the sound.

Right next to the bi-directional signal on the opposite side, a trainspotter stands at the end of the platform in a faux fur brimmed hood. The only light shining on him is the intense red of the signal bulb. There's a brief exchange of energy as I pass him. I see his face, mouth wide, smiling – just the same as I would've been. What an amazing hobby this is, and what a brilliant location for him to be standing too! Thrashing on top notch right from the emergence at the tunnel, past the end of the platform and into the darkness.

Saying goodbye to the class 37s evokes a mixture of feelings. On one hand I love them and have had a fantastic day as a passenger, a fully involved basher. On the other hand I am so glad to be one step closer to my bed. It's tipping it down, the wind creating more intense pockets of rain, and making waves of different intensity on the puddles on the platform. The 37s will head down to the sidings at Bristol Kingsland Road where they will be put to bed. We will be heading back to Gloucester where I will go to sleep in my hotel. After fifteen minutes waiting in the rain, finally signal BL 2147 changes its aspect to yellow, allowing the driver to leave. I am drenched through to the skin, but hearing the locomotives for one last time, from my usual exterior position, makes none of it matter. I watch the particles of soot shoot out from the exhaust – sheer energy. The locomotives' engines don't vibrate the ground that much, but there is a change in the air that connects me to the sensory world around me, stimulating me in a way that allows me to focus on that one thing. I can let go of everything else.

Chapter Ten ➔

The Midland Pullman

Location: Dawlish Sea Wall, 13 April 2022, 10:58

The Midland Pullman HST is my favourite passenger train on the network today. Bright blue from front to back, it pays homage to the original Midland Pullman, running from 1960 to 1966, which was liveried in the same shade of blue and hauled by two diesel-electric powercars. The modern-day Midland Pullman utilizes two class 43 HST powercars, 43046 *Geoff Drury 1930–1999 Steam Preservation and Computerised Track Recording Pioneer*, and 43055, both fitted with Paxman VP185 engines and rescued for preservation when East Midlands Railway retired their VP185 powercars. On a sunny day, the Pullman glows. Sunlight transforms the blue into an almost fluorescent shade; paired with the chortle from the VP185 while idling and the Valenta-esque scream when maxed out, chasing it makes for a great outing. Today, it's running down to Kingswear in Devon. The mainline section finishes at Paignton but from there the Dartmouth Steam Railway takes over, running down to Kingswear along one of the most beautiful stretches of railway in the UK, crossing viaducts, along beachfronts, then following the river Dart along the estuary. Goodrington Bank is a particular highlight as locomotives really have to give it some to crawl up the bank while beachgoers relax in the background, beach huts dotting the vista in an array of colours.

I had seen pictures of the Midland Pullman parked outside the paintshop in Eastleigh, and was initially put off by the bright blue. To my mind, all HSTs are meant to have a yellow front, and at first it just didn't look right without it. The blurred pictures were then aided by a few YouTube videos as it departed from Eastleigh in the dead of night on 30 October 2020, secretly making its way to its new home in Crewe. Only a few videos were caught of it, just about lit up by platform lights. The most impressive of them, the one I

rewatched the most, was taken by a trainspotting YouTuber called swearingkevo, who I assume is a man called Kevin who swears a lot. It's pouring down at Tilehurst Station as the Midland Pullman breaks into the light of the platform, cutting through the rain. Initially irritating, a GWR class 800 passes by on the fast line to London Paddington, over-powering the soundscape; however, the trailing drone of it fades and is inversely matched by the approaching rail noise from the Midland Pullman, taking centre stage. Turbochar-gers screaming, it cascades through the station just as the Great Western Paxman HSTs would have done fifteen years earlier on the same line. 'At least it sounds like a proper class 43,' I thought at the time.

The return trip from Crewe to St Pancras was its inaug-ural run, on 12 December 2020, the first preserved mainline HST back in service. I have to say, after seeing the videos of it in the light of day, albeit a drizzly winter one, my mind changed. Seeing the detailing a little closer – the glorious crest on the front, the white strip running around the driver's cab and all the way down the side of the train, the headlight recesses painted white, the words 'Midland Pullman' painted along the side of the dimpled metalwork of the powercar – it seemed so special and prestigious, in a league only with the Belmond British Pullman. I was itching to see it.

The twenty-ninth of May was the first day of 2021 that felt Mediterranean and humid in the morning. The Midland Pullman was heading down into Devon, passing through Westbury en route. I didn't fancy the station as I knew it would be crowded so I positioned myself just down the line before Fairwood Junction, about 500 metres from where Ryan first met Gordon. A footpath cuts across a field of tall, almost lime-green-coloured grass, over a turnstile – 'STOP, look, listen. BEWARE of the trains' – across the tracks and over another turnstile, into another field. The first turnstile itself is set back from the railway. It's a little walk once you've gone over it, following along tactile non-slip paving, before you actually have to STOP, look, listen and BEWARE. Standing on the turnstile, on the side of the field, was my decision as I felt I was at a suitably safe distance from the railway while also being at an elevated position so the driver would be able to see me if he wanted to give some tones. The only issue with the spot was immediately to the left: the view is interrupted by a bush. Further down the field in the direction that the service would be travelling, the hedgeline disappears, while the elevation of the field reduces. My plan to move from one spot to the other as the train passed should give two different perspectives of the powercars.

I watch back the video on my phone from that moment to reconnect with the excitement I felt. The birds are tweeting happily, very light cloud cover, the blue behind the clouds looking inviting and warm. 43055 leads under a farmer's

bridge. It's thrashing. The driver doesn't give me any tones but as he passes by I jump down to run along to the second spot. My feet land deeper and at an angle I could not have predicted. In the flurry of the moment I forgot to process how high the grass was – nearly knee height. My legs fold and I end up face-planting the ground, my camera pressed into the soil as my legs arch over my impacted upper body. It doesn't hurt too much, so I quickly get up and run down to the second spot, a bit further down. The blue carriages coast by. I look closely at the details, crests, the same as the one on the front, painted next to each slam door with retractable windows. 43046 powers by, the turbochargers whistling away, the nameplate on the side: 'Geoff Drury 1930–1999 Steam Preservation and Computerised Track Recording Pioneer'. It's a lot wider and taller than usual nameplates, to fit all the information in. The train pulls away and disappears from view. My first encounter with it was a resounding success.

The first time I met Ryan, we arranged to see the Midland Pullman together, so it has a particularly important meaning for us. Any time it's running through Gloucester, Somerset or Devon, we make a special effort to go and see it.

It has been through the wars since its inaugural run over a year ago. While preparing for a trip into Wales, 43055 was involved in a shunting incident in Eastleigh, damaging the powercar and one of the carriages. I didn't know this until I arrived at Cardiff Central Station to see it with Ryan. 'One of the Rail Charter Service's powercars is on it,' he told me just as I arrived. A freshly introduced powercar from LSL (Locomotive Services Limited) would pull the train, but it

Geoff Drury 193
Steam Preservation and Computerised Track

wouldn't be sporting the Pullman's usual dazzling blue. There was a spell last year when the Midland Pullman had to adopt yellow coupler covers while there was a dispute over requirements for the yellow panel to be fitted, even though high-intensity lights were in use. It looked totally wrong with the two bright colours contrasting on the front, distracting from the beautiful crest. While in this slightly uglier phase, it was parked up in Newport when the side of one of the carriages was graffitied by vandals. It was sad to see it like this, especially after being introduced as such a refined high-speed train. 43055 is in for an overhaul so the powercar will be replaced today by 43049, another VP185 in the Intercity Swallow livery, the same as my favourite class 43 HST powercar, 43102. Happily, the rest of the consist will be business as usual.

I know today's driver well. He's called Benji and will be taking the Midland Pullman all the way down to Kingswear. I met him for the first time on my first outing with Ryan. We were at Haresfield crossing, one of my favourites in the Gloucester area. You can stand 10 feet away from the trains blasting behind the safety of a fence, a serious rush at 70mph. Benji gave us some tones in his Voyager and later messaged me saying it was him. We then found out that he drives the Midland Pullman and is a fireman on some of the steam charters operated by LSL. We will be seeing Benji in the Midland Pullman along Dawlish Sea Wall, then following it along to Kingswear where hopefully we can get up close to 43049 and potentially have a look in the cab.

Gloucester to Dawlish takes two hours and fifteen minutes by car. For the final few miles the road opens out and you can see the sea. The tide looks to be out at the moment. It's 16°C and a bit windy, so a little too cold to swim, though I would have if it was warmer and we had the time. There seem to be sea wall reinforcement works being carried out at the moment and the usual footpath access is closed. 'I hope we can get on the wall still,' I mutter to Ryan as we swing around the roundabout. The station car park is full so we park a little further down in a diagonal bay pointing towards the train track. A class 150 accelerates out from the station in the direction of Teignmouth, rumbling by us as we get out of the car.

I have felt a bit depressed recently. Coming back from the class 37 tour, I started to overthink how my videos were being perceived in the community and rabbit-holed into thinking about reducing the content to documenting the trains by themselves without my reactions. I constantly think about how other people are feeling and how I can influence that in a positive way. Through some introspection I traced this back to trying to fit in with friends at school – noticing how people are feeling, adapting what I am saying to make them feel better or respond more positively, then gauging their reaction so I can improve my approach in the future. This made me a massive overthinker. Breaking away from those thought processes was liberating. I allowed my interests and passions to shape my identity, making me so much happier. Being in that carriage full of bashers reminded me of school, trying to fit in and trying to be accepted. I'd left that behind, or so I thought. I have reflected on the person in the corner calling me out. I can't deal with confrontation very well and to have it predictably happen in that carriage somewhat tainted my experience of the railtour, as fantastic as it was. I began to think of ways to make my videos more appreciated in the railway enthusiast community, including changing the nature of the videos I create. But no. That's the thought process that made me miserable at school, carrying me away from my own identity. Pushing those doubts aside, I watched some of my videos that make me happy and remind me why I go out to be with trains, particularly the moment when the class 37, 37884, dragging the class 365 unit was overtaken in Gloucester Station by a class 60. I was in that moment again, the pure serotonin. I'd watch it again and again, laughing every time. I want to experience a moment like that again. The sheer luck of the situation was so perfectly unpredictable, I became completely lost in the thrill of it. I am just hoping something like that will happen again. That's what would make me feel better today, a moment of unpredictability, a moment that can take me out of my muddled mind.

Dawlish Sea Wall footpath is about 4 metres wide. Heading to the east, the train line runs alongside the wall, elevated a metre and a half above the path and separated from it by a chest-high section of wall. To the right, there is a sheer drop

Britain's Most Easterly Station

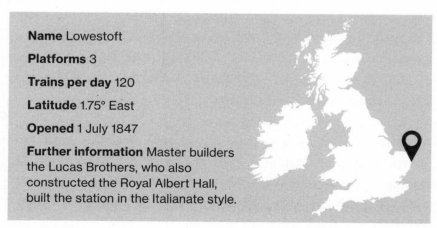

Name Lowestoft

Platforms 3

Trains per day 120

Latitude 1.75° East

Opened 1 July 1847

Further information Master builders the Lucas Brothers, who also constructed the Royal Albert Hall, built the station in the Italianate style.

into the sea, I'd say around 7 metres below. This parallelism gently curves to the right and continues for just over 1.5 kilometres where the track then swoops left towards Dawlish Warren. In the other direction, past Dawlish Station, the line runs into a tunnel, Kennaway Tunnel, then another, Coryton Tunnel, then Phillot Tunnel, Clerks Tunnel and Parsons Tunnel, then along the sea line to Teignmouth. Some of those tunnels only last a few seconds, cutting through the stacking sandstone that breaks away and undulates along the coastline. From the sea wall you can see the line as it runs between the last three tunnels, 1.7 kilometres away. Crossovers happen here all the time, usually class 150s, 166 turbos or class 800s. Very rarely does an HST crossover occur here as they run the line less frequently, but compared to the rest of the country it is one of the hot spots. Both GWR Castle Class HST and CrossCountry HSTs operate down here. The Castle Classes have four carriages and the CrossCountry ones operate seven-carriage services – as Benji calls them, proper HSTs due to them being full length.

Checking realtimetrains.co.uk (another train tracker service that helpfully displays predicted pass time as well as booked pass time, which is useful for predicting crossovers), I'm scanning to see if there are any GWR or XC HSTs due around the time the Midland Pullman passes through.

'Oh my God, Ryan, Plymouth to Edinburgh HST is due five minutes after the Pullman!'

'Crossover!' Ryan calls out.

But he's not as enthused about crossovers as I am. I'm jumping around. It'd be the perfect start to our trainspotting day out. This is just what I need. I open Traksy to see where both of them are: the Pullman is due through at 12:40, the XC HST at 12:45. The Pullman is at Cogload Junction, just outside Taunton; the XC HST has left Plymouth. From experience, it's quite a safe bet to say that the Pullman will lose time on the way. Sometimes a slower stopping service might hold it up a bit and by the look of things there's a headcode in front of it that's red, meaning it's running late and will impact the Pullman when it catches up to it. 'Yes, Ryan, look, it's got a late-running stopping service in front of it!' With the amount of line that we can see from the wall, it's bound to cross over somewhere along here!

I realize that it'd probably be a good idea to see how the signal blocks displayed on Traksy compare between when the services appear from Dawlish Warren to the east and from just before Parsons Tunnel to the west. I track a class 800 coming from the east; fifteen seconds after it passes through the signal displayed centrally at Dawlish Warren Station, it emerges from around the corner. I do the same for a service coming from the west; it emerges nine seconds after leaving the signal before Parsons Tunnel. This will have no impact on the outcome of whether a crossover will occur or not, it will just enable me to predict where exactly the crossover is going to happen on the sea wall so I can adjust accordingly.

The stopping service holding up the Midland Pullman allows it to pass through at Exeter St Davids. It is now running three minutes late, meaning there will be a two-minute difference between both HSTs at the point of Dawlish Station, close to where we are standing. I predict they will cross over more to the west. Luckily I have my binoculars into which I can put my camera lens to get a video if it's too far away. Benji has got a clear run ahead of him now. The HST from Plymouth is running on time. My heart rate is starting to increase. 'Come on, cross over, please!' I say out loud while jumping from one leg to another nervously.

Benji picks back up the time he lost behind the stopping service; he is flying along. I'm disheartened. The cloud that was hanging over me returns, and I feel disappointed. 'The crossover isn't going to happen now, Ryan.' He doesn't seem too bothered about it as seeing the Pullman with the freshly painted 43049 is exciting enough. I would've been the same but I've become too fixated on this crossover, and now it looks like it won't be happening within sight of us.

The Intercity Swallow livery powercar bombs it towards us. I hadn't noticed it approaching around the corner because I was too busy checking Traksy, hoping that maybe it'd get held up for some reason at Dawlish Warren or that the XC HST had miraculously picked up some time somewhere. I feel a rush as the Midland Pullman thumps along, but it doesn't feel pure. Benji waves at us and gives us a fantastic four-tone. I laugh and smile but it doesn't feel like it normally does. 'Look at the brilliant blue,' I say to Ryan to try and connect to the happiness of the moment. Ryan is elated, and I want to feel that way too. The Pullman continues at maximum line speed through Dawlish Station, through Kennaway Tunnel, Coryton Tunnel, Phillot Tunnel, Clerks Tunnel. I follow it with my binoculars along the stretch before Parsons Tunnel, hoping that maybe the XC HST will pop out. 43046 disappears into Parsons Tunnel. The XC HST emerges three minutes later.

Getting back into the car, I explain to Ryan that I'm feeling depressed. I want to take a moment to talk, but we need to get to Kingswear to see the Pullman arrive. The conversation diverts back to 43049.

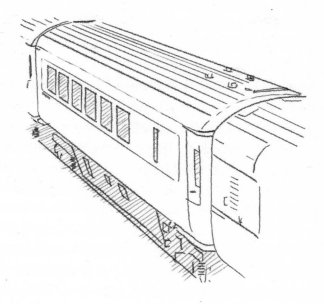

We join the A380 and pass Newton Abbot Station, which is just to the right of us. We continue along. 'OH MY GOD – look, Ryan!' The dual carriageway descends downwards, the elevation of the railway stays level, and the lined roof of the Midland Pullman comes into view. I maintain focus on the road but Ryan bounces up and down in his seat while pumping his fists. The blue looks fantastic as we dip down further – I just catch it in my peripheral vision. I'm laughing, feeling a pure boost of happiness. The Midland Pullman disappears under us as it heads down to Torquay and we continue onwards to Dartmouth. We pull off the dual carriageway and follow the traffic through the outskirts of Paignton.

Then comes the collision, the front of my Mercedes meets the back of a Land Rover with its tow bar sticking out at a closing speed of around 10mph. The impact of the tow bar over such a concentrated area pierces my bumper, bending the radiator and pushing it into the engine block. I sit there in shock. I look at Ryan.

'We're gonna need to get out,' he says.

I look straight ahead. I don't need to turn the ignition off, the engine has cut out, but I put it in park out of instinct. None of the airbags have gone off because it was such a low-speed collision. I get out to speak to the driver in front. She's fine. I see that her car has collided with the one in front. I walk round to speak to the driver of the front car – she's also fine. Ryan gets out of the car; he looks like he's in shock too. Coolant has started to leak on the road. Cars are building up behind. There's enough space to pass but it looks like I'm going to need to move the car. The two other cars involved have pulled over to the side as there's little to no damage to both of them. I put the key in the ignition to try and turn it over. The starter motor whinges but nothing happens. I manage to put the gear selector in drive and Ryan pushes from behind as I pull to the side of the road, coming to a stop outside a Spar shop. The power steering has gone so I need to pull on the wheel with maximum strength. A trail of coolant leads away from the site of the incident.

I walk back over to Ryan. 'Thanks, mate,' I say to him.

'We're gonna miss the Midland Pullman now, aren't we?' Ryan says. 'We were meant to cab it.'

'Yeah, I think you're right.'

I get the details from the drivers, a policeman assists, and eventually the two other cars involved drive away. Ryan is now crying. 'It's just our luck on the day I could've cabbed 43049 we get involved in a car crash.' I am still numb. I can't feel anything: I'm still in shock. I can hardly look at the front of the car. Recovery is coming but it will take nearly two hours according to the operator on the phone. There's nothing we can do but stand by the car and wait.

After deliberating for half an hour, Ryan decides to go to Paignton Station to catch the Midland Pullman on its return. I encourage him to go so he can salvage some enjoyment from the day, rather than stand around outside a corner shop. At least then I can experience it vicariously through him. It's a twenty-minute walk from the scene of the incident and the Pullman is due in approximately twenty-three minutes, so Ryan gets going promptly. There's a grit bin next to the entrance of the shop that looks like it will make a good seat for however long I need to wait. It's squarely corrugated with the dips around 10 centimetres deep and 20 centimetres wide. I hop on to it and the lid flexes a little under my body weight. Annoyingly, the dips do not synchronize with my hip bones so I have to rest one 10 centimetres higher than the other. It isn't comfortable.

I look at my car. Coolant is pooling underneath the twisted metal and plastic. The damage is so concentrated in the area that ate the towbar that covering it up with my hand makes the car look totally normal. It makes me sad that this could potentially be the end of the road for my Mercedes – off to the scrapyard, hopefully with most of the parts being used as spares. But I have never really felt an emotional connection to her. She doesn't have a name. She replaced the quirks and imperfections of my first car, Lucy, with efficiency and torque, which was enjoyable but didn't create any sort of bond between us.

People drive by filming me and shouting out of their windows. I try to keep my head down but every so often a 'Wuhey!' is called from a passing car or van. A family approach me and say hello, recognizing me from my videos. 'What've you been up to today?' they ask.

'Spotting the Midland Pullman. I've come from Gloucester today with my friend but we've just been involved in a car accident.' I point to my car in the lay-by over their shoulder.

They turn round and gasp, 'Oh my goodness, are you OK?'

'Yeah, I'm fine physically, it's just a bit of a shock, and people keep filming me ...' I start crying. The emotion of the situation begins to release. The family are really nice and comfort me before asking if I want anything to drink. I politely decline. After they leave I carry on crying quietly, thinking about the collision, the shock. Why am I weeping? It keeps coming and coming like the outlets of a dam being opened into the valley below. The car is almost certainly a write-off. I think about the people whose day has been inconvenienced by the collision, but they have all been lovely. I can block out the people jeering out of their windows. Missing the Midland Pullman is disappointing but I can always see it another day. Ryan crying because of the crash – that touches me the most, because I know it ruined his day. But I also know we will have many more great outings in the future. I try to process the whole situation. Earlier I was so disappointed about the crossover not happening, I had built it up so much in my mind. When that moment didn't happen, the whole construction of the situation fell to the ground and the emotions that I attached to it crumbled in the same way. I felt hollow and disappointed. I think that might be what is pouring out of me now.

Trainspotting is about being in the moment. Everything going on in my mind – ideas, relationships, memories, feelings, actions whizzing around the inside of my head like daredevil motorcyclists going round and round the wall of death, narrowly avoiding one another – all comes to a stop and settles in the middle. Trainspotting focuses my brain on one thing and corrals everything else for a moment. It clears away the detritus left behind from cluttered thoughts and feelings. After a day of trainspotting, I am tired but I feel happy and fresh. I imagine it's similar to casting a line into the water and waiting for a fish, or cycling for miles, focusing on each stroke of the pedal. Coming away from it, the endorphins keep me buzzing. That's why I come back to it – it makes me feel happy. I realize now, after the disappointment of a day that didn't go to plan, that I am trying to make these unpredictable situations happen, and that is a fool's errand. The best moments arise out of pure chance; that's what makes them so amazing. To try and construct a euphoric trainspotting moment is like filling a pond full of big fish: you are more likely to catch one but the sheer surprise and the endorphin release that follows will never be there. Worse still, when a big fish isn't caught, or the trains don't cross over, the disappointment will be more intense in a situation that could have been just a normal day. Happiness lies in the moment, focusing on the trains and letting the infinite combinations of life unravel.

Location: Paignton Station, 13 March 2022, 17:54

Ryan is waiting for me at the station after seeing the Midland Pullman on its return journey. It stopped for a few minutes and Ryan explained to Benji what had happened. He kept me up to date with all the happenings via Instagram messenger. We will be getting a class 150 together back to Exeter St

Davids where Ryan will luckily be catching a Castle Class HST back to Gloucester and I will be getting a GWR class 800 bound for London Paddington, getting off at Westbury. With my car loaded on a tow truck, a kind man has brought me in his double glazing van to Paignton Station. He watches my videos and seems like a nice person. With only a few minutes until departure, I jump out of the van just outside the station and give the man a £20 note for his random act of kindness. Ryan is waiting at the end of the platform, beckoning me to hurry. I jump on board the recently refurbished class 150, with exactly the same interior as the purple and pink 150 we saw at Bristol Temple Meads before the class 37 railtour, just with a different livery to remain in keeping with the company's dark green aesthetic.

We sit down at a table facing each other. Ryan seems his normal self, which makes me feel better. He is still buzzing from the Midland Pullman departure. Benji did some good tones and the locos were at a higher notch than usual due to the slight incline heading east from Paignton. He shows me the video then AirDrops it to my phone so I can watch it back later too.

The class 150 takes us up the incline from Paignton along to Torquay, which looks like the English version of Monaco. The weather has really taken a positive turn, the clouds dissipating more and the six o'clock sun giving a greener hue to the sky on the horizon. The white tower block buildings contrasted with the quaint seafront and Mediterranean-looking water with little boats bobbing around fascinates me. In all my researching of trainspotting locations and general geography in the UK, I have never realized how Torquay looks as if it exists on the French Riviera. At least from the view from the train. We pass the location where the A380 runs just above the elevations of the tracks, where we ran along the Midland Pullman earlier, and through Newton Abbot Station. The river Teign passes beneath us and we swoop around to run alongside it. Calm and wide, it looks to be the perfect backdrop to a trainspotting location. The tracks run so close to its edge, perhaps the other side would be better; however there wouldn't be any foot crossings as the other side would drop straight into the water.

'Ryan, imagine if we got a boat and took it up the river right next to the tracks?'

'Hee-hee, that would be funny.'

I start to think about how it could be done as Teignmouth approaches. Small boats populate the water and families seem to be packing up in the car park after a day by the beach.

We connect with the water again after Teignmouth. The sea is calm but wrinkled and creased like wrapping paper that has been scrunched up tightly then flattened again. It's hard to judge the scale of the sea from this position as the track-side wall hides comparable objects like rocks and pebbles. We are travelling at 50mph now, according to my internal speedometer. The uniform concrete wall suddenly chops into a stone wall, rising up and down with the inconsistencies of the rock sizes. It probably seems level when stationary but passing it at speed reveals the rustic nature, far nicer and more in keeping with the landscape than the concrete one. The track starts to pull away from the water and some of the beach comes into view, waves crashing gently on the sand, leading outward towards the approaching cliff face. This must be the first of the tunnels heading in the direction of Dawlish, Parsons Tunnel. Red sandstone, crumbled and ridged, it looks as though it could be reduced to dust at the lightest abrasion. It protrudes into the sea by 80 metres, at a height of around 15 metres, the tunnel cutting straight through it. A small stack stands further out at sea, as though someone has kicked a chunk of rock off the top of the cliff.

DOO DIII – the driver lets off a two-tone. 'Tones!' Ryan pipes up, then looks back down at his phone. Is he passing a whistle board or toning for people? I wonder. Three people flash by. I catch a brief glimpse of them walking along the sea wall: a lady with brown hair swept downwards across her face in the wind, looking straight ahead but with an arm raised, waving at us as we pass by; a mother holding the hand of her son, her other hand raised in the air, waving. She has a big smile across her face while looking down at her son whose mouth is agape, stunned by the tones, eyes fixed on the passing class 150. I laugh. I know exactly how he feels and feel a smidge of the same energy in that moment. We pass into Parsons Tunnel. The whirring of the wheels reverberates in the enclosed space, wind rushes through the windows under the increased pressure. Emerging on the other side, there's Dawlish. The track banks to the left, opening out the view. I sit back into my seat and relax, connecting with my senses and my feelings. The happiness returns.

Chapter Eleven ➔

The English Riviera

Location: Paignton Station, 21 May 2022, 13:15

Just over a month since the car crash. I still think about the moment it happened and my feelings before and after. The train journey along the English Riviera that followed reminded me of why I love trains and trainspotting. It's an escape, a time to turn off the mind. I stepped off that class 150 feeling like the cloud that was over my head had released its rain and evaporated. Better still, Lucy is now my full-time trainspotting travel companion. She struggles more on the hills, needs a good booting to accelerate on to the motorway, but loves navigating the country lanes that trainspotting leads us along. I have to press the lock button nearly twenty times before she finally complies, which is something I did not miss, but driving her makes me feel connected to our journeys in a way that I have missed.

We have arrived in Paignton successfully. We are setting out to do what we failed to do last month. The Midland Pullman is coming back! This time it is not going as far as Kingswear, just stopping at Paignton, but Benji will be at the helm again so hopefully we will get a chance to go into the cab if there's time. The weather is a lot nicer today, so the blue of the Pullman is bound to look glorious. 43046 will be leading this time unlike before when 43049, the Intercity Swallow-liveried powercar, was at the front. It will be the trailing powercar for this leg of the journey, which is good news for us as we were worried the 43055 would be back in service, making it unlikely to see 43049 on the Midland Pullman again.

Over the footbridge, we make our way to platform 1 where the train will be arriving. It is so glaringly bright that I have to clench my eyes together to ward off the intensity of the sunlight. A class 166 sits on platform 2, a mundane-looking unit usually but in this light the yellow and the dark green look stunning. The Paignton depot and the station for the Dartmouth Steam Railway are situated behind us. I peer over the fence to have a look at their shunter numbered

D3014, chevrons contrasting up the rear – a feature that I love so much and know so well from when I was younger, always a familiar sight around Willesden TMD.

The D3014 numbering is something that snags in my mind a little bit when it comes to diesel locomotive appreciation. Class 08, class 31, class 33, class 37, class 43 ... I can put a face to the name, but pre-TOPS (Total Operations Processing System) numbering, locomotives would be given a letter followed by up to four numbers. For example, diesel locomotives, such as the class 08 shunter, as I know it, would be given the letter D, for diesel, then an identification number such as 3014. E was for electric locomotives. Steam locomotives didn't have a letter at the beginning, just up to five numbers instead. TOPS came in to improve the control and planning of freight and required locomotives to be renumbered with a five-digit number so they were fit for the processing systems used. Each locomotive type was designated a class, identified by the first two digits; iterations of that class came in the third digit, and the last two digits were used to identify the specific locomotive. For example, 37510: class 37, subclass /5 (featuring a newer alternator instead of an older English Electric generator), loco number 10 within the sub-class. This is the identification of locomotives that I am familiar with, so pre-TOPS numbered locomotives don't appeal to me in the same way. I find pleasure in categorizing things, and in systems that catalogue and create order. Sometimes heritage groups will give a locomotive a fresh lick of paint, returning it to British Rail green and assigning it its pre-TOPS number, which I find a little dull. In the case of the 08 next to me it was actually one of the first to be built and wears the same designation it was given back in 1952. Controversially, I think it would be far more interesting if it was TOPS numbered and given an EWS livery or something similar.

43046 rumbles into platform 1 with the level crossing chiming away, gates down, preventing the seaside enjoyers from continuing, unless they want to scale the footbridge, which for some seems the less desirable choice. I wonder how many will stay around to see the train pass by – well worth it this time if so! 43049 is on the back so we watch it coast in from midway down the platform to see it once it's come to a stop. It's awfully crowded down by the end of the platform near the level crossing and our videos will likely be inter-rupted. It's a lot better midway. The other end where the lead powercar will halt is packed with enthusiasts.

Today the train is shining the most luminous blue, as bright as I have ever seen it. Benji pulls it in slowly in his black waistcoat, in keeping with the original outfits worn by the staff on the Pullman. Benji is a nice, friendly man and seems to take his job very seriously, confirmed by the fact that he drives the Pullman routinely in the south-west. The clatterly idle of the VP185 passes us and we start to make our way towards the crossing to see 43049 up close for the first time.

'It's too long,' Ryan grumbles as the powercar stops just short of the edge of the platform.

'Come down here, Ryan – look.'

I make my way down the slope to street level where non-train enthusiasts are puzzled by the bright blue carriages and Intercity Swallow-liveried powercar. Some even stop halfway up the stairs and turn round. There's a shoulder-height three-spike fence that separates the footpath from the tracks – an easy rest for the wrists to balance on in between the spikes for a steady shot. The red plaque looks fantastic:

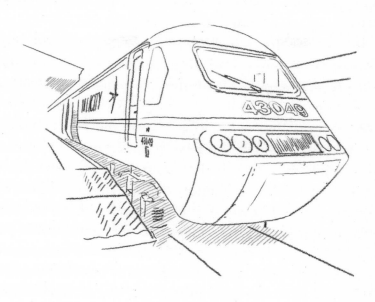

'Neville Hill', small and discreet, sitting in the same position, just behind the side grille, as I remember it did when it was with East Midlands Railway, at a time when I was riding behind this exact locomotive on normal passenger services. I notice the NL sticker near the driver's door too. It indicates the depot the locomotive sleeps at during the night, Neville Hill (43049 used to sleep at Neville Hill depot, hence the nameplate). Her livery is a nod to her life in the past and the depot sticker is a nice touch too. She now sleeps along with her other preserved mates at Locomotive Services Limited's depot in Crewe, most of which, HST-wise, would also have lived in Neville Hill back when they were on passenger services. The heritage livery looks fantastically fresh, shock absorbers painted white and the steps and springs painted yellow – delicious detail to see on 43049. The refurbishment reminds me of how the industry is full of enthusiasts who have the power to make choices to do with the trains that send ripples of pleasure through the community. Unable to get any closer, Ryan and I decide to leave 43049, slightly unsatisfied with our shots, to go and see the Midland Pullman depart from the front, hopefully with some thrash, tones and clag.

Typical jostling ensues for a good spot. Ryan and I are a bit late to the show and have to make do with the second row. Luckily, at 6 feet 3 inches I can raise my arms above the first row and get a reasonably decent, uninterrupted video. The only issue is I can't see the screen so I have to trust my brain's motor judgement to position my wrists at the perfect angle. Annoyingly, the front of the locomotive has passed the end of the platform, again, in order to fit all of the coaches on. The VP185 rumbles away next to us, the side grille offering angled growls from the engine inside.

There's something so satisfying about the grille – a square taken out from the imperfect metalwork running along the side of the locomotive and replaced with slats of metal tilted down at 45 degrees. Reinforcement bars run behind the slats, from the bottom left to the upper centre, then straight down to the lower centre, then up to the top right, like a saw tooth. At too high an angle, the reinforcement bars cannot be seen, but usually you can see them peeping through. Oddly, there is a cut-out in one of the slats near the bottom right that I notice sometimes with class 43s, not featured on all of the locos but only some. Perhaps it was a modification made to later sub-classes. The slats take a break for around 20 centimetres before the roofline, then continue across the roof like gills. I really find them so satisfying.

Benji powers up. The VP185's gargly rattle doesn't reach a high enough rpm for the turbocharger to spool up sufficiently to hear it. Light clag floats out from the top of the exhaust. Reflections in the metalwork of the powercar augment and sway as the Midland Pullman starts to accelerate. There is no thrashing as the consist needs to negotiate a succession of tight radius curves leaving the station due to the two tracks from the two platforms converging just before the level crossing, where the tracks coming from the Dartmouth Steam Railway depot and station converge too, to form the second track that passes over the crossing. 43049 is idling. It is fantastic to see the train but the videos and photos we get are mediocre and there was certainly no chance to cab either of the powercars. We didn't even get to say hello to Benji other than the wave he gave us from the cab. I don't feel emotionally caught up in it whatsoever. It is not quite the encounter we wanted but I'm holding on to the positives from it. I'm also looking forward to seeing the return journey. We are going to get a boat and return to the river Teign where it runs alongside the tracks. The location I spotted last time. It's meant to be.

We return to the car park with the aim of finding where the Midland Pullman has parked up. Lucy opens after one click of the unlock button this time, rather satisfyingly. The interior isn't in the best shape as I used to transport my mountain bike around in her, bashing the B pillar and muddying the panels as I negotiated it into the boot. There are a few bits of litter lying in the footwell and Ryan does not hesitate to point out that it's less clean than the Mercedes.

'I miss your Mercedes, it was so elegant,' he reminds me.

'Yep, well, we're going to have to make do with Lucy I'm afraid.'

We pull back on to the road that runs in the same direction as the tracks but I can see it deviates to the right as we head west. We take the next left at a T-junction on to the A379, which seems promising as it appears to be running back towards the direction of the track. I love maps and following directions but I also love utilizing my internal compass to guide me through unfamiliar roads, as I have decided to do now. We pass another left turning – I suspect this is the road that has the level crossing. 'Yep, look, Ryan, there are the tracks.' I catch a glimpse of the bright red level crossing in my peripheral vision. Earlier, from the edge of the platform, it looked ancient, almost as though it still needed to be operated by hand when it was closed. After seeing it rise into the air automatically, I am proved wrong. It is also funny to see how knackered the left-hand side track is in comparison to the right-hand side, heading west. I'm assuming it's more knackered but it might actually only be a matter of appearance due to the soot and grime left behind by the steam locomotives heading up to Dartmouth and back.

We increase in elevation as a roadbridge passing over the tracks approaches from the left. As we near the crest of the hill we spot the Pullman. 'There it is!' Ryan shouts. I smile as I navigate a mini roundabout, carrying straight on, running parallel to the tracks. I take the first left to see if that gets us any closer. We pass along an interconnecting road that leads us to another, very quiet street that runs in between the A379 and the tracks. It's bordered by houses on one side

and a high stone wall on the other, shielding the railway from view. NO BALL GAMES signs are knocked up along the wall every 10 metres. The road is quiet because it's a dead end. As a child, this would've been the perfect playground. I can just see myself setting up a little portable ramp and flying off it on my scooter and BMX bike. Leafy tendrils reach over the top of the wall, in some cases touching the ground and providing the scaffolding for a flow of leaves to engulf any sign of breezeblocks. Bins line the wall. There's a gap in the structure, covered by a temporary metal fence.

'It's there, look!' 43046 sits clagging away at a standstill. I am astonished at the number of tracks in the sidings here. For a town that has such light rail traffic, a width of six tracks seems excessive. The Pullman sits there on its own, basking in the sunlight. 'This is perfect light to get a good shot of 43049, Ryan,' I say, assuming the road leads us to a clearing. There should be a good line of sight from there.

Britain's Most Westerly Station

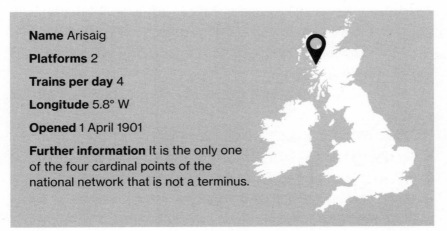

Name Arisaig

Platforms 2

Trains per day 4

Longitude 5.8° W

Opened 1 April 1901

Further information It is the only one of the four cardinal points of the national network that is not a terminus.

We park at the end of the cul-de-sac with a view of the roundabout we passed earlier, only divided by a patch of grass and a rusty cross-linked fence. Not a clearing in sight. 'This is bad here,' Ryan mumbles. We get out and examine. Ryan goes for a wee in the dense bushes hiding 43049 from us while I pull myself up on to a clear patch of wall, just high enough so my eyes can see over. I scoff quietly. The sun, just behind my shoulders, is lighting 43049 perfectly with two tracks clear on the near side and three tracks behind, interrupted only by the occasional jogger over on the foot-path opposite.

'It's fantastic!' I call over to Ryan as he walks over.

'Eh?' He looks puzzled. I don't think he saw me pull myself up for a peep.

'Look, come with me.'

We walk down towards the spot, where I realize that I will be able to get myself to a better elevation if I stand on one of the industrial bins lined up along the wall. We find a sturdy-looking Biffa bin with all four wheels pointing in different directions, meaning it'd be hard to shift. 'Easy now,' Ryan advises. The black bin lid is hot to the touch after being blasted by the sun's intense infrared radiation. I hoist myself up, tentative on the flexible lid, and peep over.

'Phwoah, this is fantastic!' I immediately get my camera out to snap a few shots.

'Take some for me as well,' Ryan calls up, handing me his phone.

Making sure I get the exposure right, I take a few, the reflection from the sun making it hard to see the screen. 'I think that's all right,' I say as I pass the phone back down to Ryan, peering up at the windows of the houses to see if anyone has seen me clambering over the bins.

Clap. My shoes slap against the hot tarmac as I jump down.

'Couldn't do that, my knees would go,' Ryan laughs.

I take a look at the photos in the shade. 'Finally, a good picture of 43049.' Ryan seems a little bit miffed that he couldn't take it himself as we move back to the car.

The Midland Pullman will sit for a few hours in the sidings, enough time for us to head to Teignmouth to rent a boat. I was worried that Ryan wouldn't be on board with the idea as he can't swim and I imagined the open water might be too scary for him, but he seems really excited. I messaged Benji to tell him we are going to see him just before Teignmouth Station so, if he catches a glimpse of us, he might give us some tones. I told Ryan about Benji's acknowledgement of us being in a boat for the return journey and the prospect of tones, and this made him even more willing to give it a go.

Teign Boat Hire seems to be just next to the beach car park, so I put it into Google Maps and we head off. Ryan examines the schedule through Teignmouth Station: if we are lucky we might be able to have a test run with a CrossCountry HST heading towards Plymouth. The fantastic part about using the boat is that we can technically find the perfect spot for either direction; if there are any foliage obstructions, we can simply motor along a little more. My only worry is about standing up in the boat and waving at Benji as he blasts by and the stability issues that might entail, particularly with both me and Ryan not being sportsmen or people who require good balance on a routine basis. But we will have lifejackets, and the topography of the river suggests it is shallow at this point, particularly at the sides. If we fall in, it'd be rotten, but I don't think there would be any danger to life.

Location: Teignmouth, 21 May 2022, 15:22

The car park at Teignmouth Beach is busy. It is a warm afternoon with the sun beating down, a clear invitation for a sea swim, but today I have a different motive. The car park is another 'day tariff only' – £6.40, almost GWR station prices! I walk over to the ticket machine and a young couple, about the same age as me, pull up and the passenger-side window rolls down. A girl in sunglasses waves at me to get my attention.

'Excuse me, are you about to buy a parking ticket?'

'Yes, I am.'

'Here, take ours, it's got till the end of the day on it.'

'Oh wow, thank you very much. Are you sure the ticket isn't assigned to your number plate?'

'No, I don't think so.'

She hands it to me and I check to see if there's any link to a number plate – nothing.

'Thank you very much, have a wonderful afternoon!'

They drive off.

'Someone just gave me their ticket, Ryan!'

'I know, I heard.'

I slide the ticket on to the dashboard. Lucy locks hastily and we make our way towards the beach huts on Teignmouth Back Beach, the other side of the spit of land that brings the river around to the right before being freed into the sea. Presumably the water isn't very nice to swim in on this side, more sewage potential and river critters, although I imagine the depth of the river here is determined by the tide, so when it's high, it's probably as good for swimming as the other side, in the open ocean, probably warmer too.

We near the edge of the car park and the river comes into view. 'It's a lot lower than I remember it, Ryan.' Quite a few of the boats that I saw bobbing up and down in the water when I went by on the train are now tilted over, lying on algae-covered silt. There is a passage of water that runs along the centre of the estuary but it doesn't go anywhere near the side.

Ryan shakes his head. 'This ain't good.'

We walk down to the beach huts. The concrete stairs down from the car park lead us around the back of the huts where it smells like urine. There's a gap between two of the huts and we walk through. I look out to the railway line and where it follows along the river, momentarily interrupted by the A379 which bridges across the water. The boundaries of the river near the track seem to be touched in places by the water – promising. The Teign Boat Hire hut stands to the right of us. A man in his late fifties sits inside.

'Excuse me, would it be possible to hire a boat today?' I ask him.

He looks up and walks towards me. 'Certainly, how long are you looking to rent the boat for?'

'Three hours at the most,' I reply.

'It's £40 an hour so that'll be £120.'

'OK, that's fine.' I turn round to look at the river. 'Just wondering, do you think there's enough depth in the water out there to take the boat near to the train tracks?'

'Oh no. No way. You'll get it stuck then have to wait till the tide comes up again.' He looks at his watch. 'Which will be in six hours.'

I look at Ryan with a disappointed frown. He shakes his head but in an understanding way.

'OK, thanks, I don't think we'll do it today but we may come back another day soon.'

'No worries, thanks!'

The man turns round and walks back into the hut, we walk back to the car.

'We're gonna have to catch it somewhere else,' Ryan sighs as we sit in the car with the windows open to let the breeze blow the baking air out from the interior.

The disappointment is a lot stronger with the tides not working in our favour. It was the experiment I was looking forward to all day and it was supposed to be the saviour of the uneventful morning in Paignton. Trying not to hold on to the feeling, I check live.rail-record and see if there's anywhere the Pullman will stop, hopefully for long enough to give us a faint chance of cabbing it. Exeter St Davids for twelve minutes. 'That's perfect, Ryan, look.' A microdose of serotonin creeps back into my body. The possibility of getting up close with the Pullman again at a location where both powercars will be on the platform, and getting to see Benji again, makes me feel like there's hope for the day.

Ryan searches on realtimetrains for the Pullman's journey back through Exeter. 'Oh my God, XC HST is gonna be in at the same time.'

'No way! From Plymouth or to Plymouth?' I ask, considering in my head if it'd be possible to catch a crossover video of them at speed – superior to when they're just in the station, which is a more likely occurrence.

'Coming from Plymouth,' Ryan states.

'No chance of a crossover then.' They're both coming from

the same direction so it'd be impossible to get a crossover shot. In fact, the reason why the Pullman is waiting twelve minutes at the platform is to let the XC HST in front of it. It feels like a breath of fresh air anyway. 'Two proper HSTs, just like Benji would say,' I laugh. The XC HSTs are composed of seven carriages, the Midland Pullman is composed of eight at the moment; it'd usually be nine but one of the carriages is at Crewe for repairs.

'Ah, only problem is that they're only at the platform together for three minutes,' Ryan points out.

'That'll be all right. The Midland Pullman has to let it through so they're guaranteed to be at the platform at the same time.'

'Fantastic!'

Just as we reverse to pull out I remember the parking ticket that was given to me. 'I'm just going to find someone to give this to, one sec.' I edge the car forward into the parking space again and get out. A family in a Subaru Impreza has just pulled in a few spaces away and the driver is getting out. I can hear a baby crying from inside the car.

'Excuse me, would you like this day-ticket for the parking? It's not assigned to a number plate or anything.'

'Yeah, sure, thanks.' The man takes the ticket, and I walk away. He was slightly less grateful than I would've expected him to be but he might just have been through an awful journey full of crying, or maybe he's just not an emotionally expressive type of person. Either way, it feels good to be passing on the random act of kindness.

Pulling out from the car park, Ryan and I talk about passing on kindness.

'I love it,' Ryan chirps. 'Karma works in mysterious ways. I really do believe in karma.'

'Yeah, well, I think if there's any opportunity that positive energy can be put out in the world, it should be taken. Just like our videos. Gordon does something kind and gives us a tone, we enjoy it and make a video, then that positive energy is amplified and distributed, sometimes across the world.'

'All it takes is for a driver to want to make our day better, then he gets good karma.'

'Exactly,' I reply.

'We make drivers happy too, like when they see us at a bridge and see us smiling, it makes them smile and feel good too, especially when they know they've made us happy.'

'Yeah, I can imagine it might get a bit lonely sometimes being a driver.'

Ryan looks down at his phone. A thought suddenly pops into my head. I wonder if any Castle Class 43s are around in Exeter when the Pullman is running through. It's not impossible, though three HSTs from different train operating companies at once is extremely rare. Improbable. I ask Ryan anyway: 'Can you check to see if any Castle Classes are in at the same time?' I look left and then look right before pulling out from the car park.

'Oh … oh my God, there is!' Ryan shouts. 'It comes in at 17:19 and leaves at 17:24 on platform 4!'

I laugh in disbelief and gasp at the thought. I haven't seen three HSTs lined up together since the good old days at Paddington Station where their sharply elegant noses would reflect along each bay like opposing mirrors. I can't believe it. I focus on driving while thinking in the back of my head about how it's going to work.

'Wait … so what time is the Pullman due in and out?'

'17:20 and 17:32 on platform 6.'

'OK, what about the CrossCountry HST?'

'Erm ... 17:23 and out at 17:27 on platform 5.'

'Holy shit, that means they're going to be lined up on platforms next to one another!'

'Yep. This is gonna be epic.'

While on the twenty-seven-minute drive, I give some thought to how the situation will unfold, running it through in my head. So the Castle Class is first on platform 4. It can't leave early otherwise it won't intersect with the CrossCountry HST. The Midland Pullman we can rely on as it will be sitting at the platform for the CrossCountry HST to pass, it just needs to stick to time so it doesn't miss the Castle Class. The Castle Class is heading to Penzance so it will be at the south end of the platform but the other two will be at the north end. This could cause a problem because of the Castle Class not being full length: the powercars won't line up. Unless I head down to the south end and get the rear powercars of the Midland Pullman and the XC HST with the lead powercar of the Castle Class set. That will have to be it. But I want to catch the Pullman pulling in from the front so I can say hello to Benji.

After my thinking session, I'm ready to propose my plan to Ryan. 'OK, here's what I'm going to do. If all runs on time, I'm going to watch the Castle Class pull in from the north end and it will then come to a stop at the south end. Then I'm going to see the Midland Pullman come in from the other direction but with 43049 coming to a stop next to us at the nor—'

'Wait, 43049 will be leading.'

'Yeah, 43046 led into Paignton earlier.'

'Awesome, we might get to cab it after all!'

'Yeah! Anyway, once the Midland Pullman has come to a stop, I'm going to say hello to Benji then make my way down to the south end of the platform while, hopefully, the XC HST pulls into platform 5. Once I'm at the bottom, I should have the blue Midland Pullman powercar, XC powercar and GWR powercar all in the same place, lined up for a photo.'

'What am I gonna do about cabbing 43049?'

'Sorry, mate, I'm sure you can do it yourself. Maybe Benji can take a photo of you in the cab?'

I feel bad that I'm not able to assist Ryan with his shot but I really cannot miss the opportunity to catch all three HSTs at once – something I may never see again.

Location: Exeter St Davids Station, 21 May 2022, 16:23

We arrive into Exeter St Davids Station. There's a nice mix of passenger trains here, arguably one of the best in the region. SWR class 159s, GWR 166s, 165s, 158s and 150s, not to mention the CrossCountry and GWR Castle Class HSTs. There's a South Western Railway class 159 on platform 1 as we walk in; the first unit is in the orange, red, yellow and white livery, the second unit is in the grey, dark grey and blue livery. I see these often around south-west London, blasting to and from London Waterloo. It makes a nice change to the totally electrified third-rail network in the southern region to have a DMU at top notch rattling through the suburban stations. Funny to see it out here at the other end of its journey.

We walk to the fly-spattered blunt nose of the lead unit to watch it depart. Every square centimetre of the front of this 159 is plastered with flies. After its thrashy departure, we make our way to the stairs leading over the tracks. Each step flexes very slightly under the foot; a grippy covering with yellow edges improves surface friction over what would've been plain wood. The stair banister is wide, too wide to close a whole hand around, and painted a deep green. Smaller-radius yellow handrails jut out by way of replacement and in accordance with modern safety standards. The layers of paint on the banisters are evident as the fine creases of the original woodwork have been filled in by the thick paint, leaving rounded approximations of the old detail. The roof-line of the walkway sits perpendicular to the canopies of the station platforms. It's covered in thick grime: rain has landed carrying particles of ash and dumped them in streaky lines towards the guttering. Light is left weak and slightly yellow due to the UV-degraded plastic. Ryan walks on ahead while I

examine the walkway. He heads towards the platform 5 and platform 6 signs at the end with arrows pointing left and right for 5 and 6 respectively, even though they end up at the same centre point of the island platform. It's fairly empty on the platform with only a few spotters dotted around. 17:01 reads the platform departure board. We walk to the north end of the platform and I recap my plan, ready for the arrivals.

Just visible under the footbridge, around 350 metres away, the yellow top and bottom of 43049 *Neville Hill* comes into view, leading the Pullman, crossing the river Exe on the way into the station. Simultaneously, two distinct tones from a class 43 echo behind me from the north side.

'That must be the Castle Class, Ryan!'

He giggles. 'They were epic tones!' I presume they were for workers by the tracks up the line.

The Midland Pullman seems to be crawling into platform 6: perhaps the signalling isn't in Benji's favour. I turn round to see 43198 *Driver Stan Martin & Driver Brian Cooper* coasting at a decent rate into platform 4, a lot faster than I was expecting, remembering that it needs to bring all four carriages to the other end of the station. 43049 and its bright blue coaches have made their way on to the platform by the time the rear powercar of the Castle Class passes me and quickly disappears behind the waiting rooms due to the curvature of the track. Platform 4 is dead straight, while platform 6 swoops away from the linearity of platforms 4 and 5, then converges back once past the waiting rooms and other miscellaneous rooms to be parallel with 4 and 5 towards the end of the platform where I'm observing from now. This creates a fantastic picture as 43049, the lead power-car of the Pullman, re-emerges while the rear powercar of the Pullman, 43046 *Geoff Drury 1930–1999*, draws in to the beginning of the platform, marking the end of the luminous blue snake. Meanwhile 43198 and 43192 *Trematon Castle* have uneventfully come to a halt at platform 4. A frame in the video has all four powercars in it, in varying degrees of detail; what makes it even better is that it's just as 43049 passes into the light and is perfectly in line with 43192. The rattle from the 2,250hp Paxman VP185 makes itself clear as it approaches us. Considering the rear powercar has long

passed the southern end of the platform I'm hoping Benji will stop before the northern end, so we can get some photos of it closer up, as we were hoping to do at Paignton earlier.

There is a peppering of trainspotters along the platform now, some with their parents, teenagers in groups and older spotters. A concentration of them stand at the end of platforms 5/6. Usually it's a good indicator of where the train will come to a stop if a spotter is unfamiliar with the station: look to where the locals position themselves.

'He's not stopping here, Ryan, I think he's going up to the end.' Benji isn't decelerating as fast as I'd have hoped.

We start walking along as the idling rumble of the engine passes us. My mouth goes into a poked-out whistling position in reaction to the sound. It never gets old, such a beastly sound. Benji briefly overtakes us but starts to decelerate a little more. 'Oh!' Ryan shouts. The Midland Pullman comes to a stop equidistantly between where Ryan and I were originally standing and where the other spotters are at the end of the platform. We all swarm to the locomotive, staying behind the yellow line and commencing our ritual of photos from different angles.

Benji gets up from his seat after safely resting the train and immediately gets out of the cab. 'I'll be two seconds,' he says as he locks the cab door and starts walking down the train. He is very good at his job and will only give time to us spotters once everything on his to-do list is complete, and if there's time. In some cases, that means just taking off right away once all is done. There's always time for a wave and a smile from him though.

After taking in all the details of the freshly painted locomotive with my eyes, I pick out every detail with my camera. Fellow spotters have distributed really nicely around the locomotive and we almost take it in turns to photograph certain parts. Chuffed is the word – it all seems to have come together! The train that we failed to see a month ago after the crash, failed to get a good photo of in Paignton Station earlier, and failed to get a boat to be close to in the water, is now rumbling next to us. I can touch it. I can feel the rumble in my fingers and hand. Rumbling that had only before touched me through vibrating air particles.

I step back to check on the Castle Class. 'Oh my God, the XC HST is here!' I shout as the seven-carriage high-speed train cruises into platform 5, its sibling on either side. 'OK, it has happened,' I think to myself. I take a moment to process what is actually happening then I walk towards Ryan and the other spotters. They might not have heard me over the Paxman clatter. 'The XC HST is pulling in as well,' I tell the small crowd of fifteen spotters.

'No way!' ... 'What're the chances?' ... 'Oh, amazing!' Even the older spotters seem to be excited.

'Castle Class stay there!' I shout. The spotters laugh. It feels nice that there's a shared sense of happiness and that other people are being vocal about it.

The lead powercar of the CrossCountry set is approaching. I suddenly realize that, in my plan, I was meant to be down at the other end of the platform by now. 'Ryan, stay there, I'll be back.' I start fast-walking down the platform as 43301 burbles by. I want Ryan to stay by 43049 just in case Benji comes back before me and offers a moment in the cab. I'd hate it if we were to go down to the south end of the platform, not get the triple shot, and for us then to miss an opportunity to get in the cab too. I pass the centre of the platform. Benji is walking back towards the lead powercar.

'Three HSTs!' I shout to him as I walk by.

'Yep!' he calls back with a smile.

I hear the dreaded whistle on platform 4. I break into a faster fast walk. The Castle Class HST has been dispatched and will be building up power any second now. I pass the end of the last building on the platform. 'No, don't go!' 43198 starts to move off from what would have been the perfect shot – all three powercars lined up together. I break into a run to get ahead. I turn round and with my camera recording catch all three class 43 HST powercars lined up next to each other, 43198 edging away. I fist-pump the air in triumph. It's happened. It's a different sense of joy from the feeling I get from a thrashing class 37. It's lacking the goosebumps and overwhelming electricity in my body, but the serotonin and adrenalin are pumping through.

43301 leaves next on its journey to Scotland, MTU rumbling, leaving the Midland Pullman on platform 6. Benji gestures to us that it's OK if we want to come up and have a look in the cab. Ryan goes first. He sits in the driver's seat and sticks his thumbs up as he leans closer to the windscreen while I take photos of him. We switch, and I take a seat at the helm, hearing the VP185 rumbling over my shoulder. I am absolutely buzzing but keep my time in the cab very brief as I know Benji is set to leave in a few minutes.

'That was fantastic, thank you! I hope you have a nice rest of your journey.'

'Yes, nice to see you again,' Benji replies.

'Nice one, Benji, safe trip.'

'Take care, see you in a bit.'

Benji departs with a two-tone and a smiley wave. The power builds. Clag pours out from the exhaust. The coaches whoosh. I am totally happy, and glad to see that Ryan is beaming from ear to ear too. The day has gone in a different direction from the one I had originally planned, just the same as it did a month earlier when we were looking to cab 43049 at Kingswear. This time the turning cogs, gears and wheels that can determine the outcome of each individual moment have given us a combination that strayed us away from our plan but thrust us back towards what we originally set out to do. It's this unpredictability that gives trainspotting its thrill, and gives me a reason to keep coming back again and again.

Chapter Twelve ➜

Cambrian Bliss

Location: Shrewsbury Station, 22 May 2022, 09:56

Shrewsbury Station car park is significantly smaller than I expected it to be, considering the number of trains passing through and that it's one of the main arteries leading into Wales. A triangular cut-out from the surrounding redbrick buildings, the car park lies within, bordered by the dramatic limestone station building and the walls of Shrewsbury Castle. A few of the spaces are taken up by British Transport Police vehicles, the rest are jammed with normal civilian cars. There's one space that miraculously seems to have been freed. Lucy makes a dart for it and we settle for a moment once parked.

Today I will be chasing a Pathfinders railtour from Shrewsbury to Pwllheli, in north-west Wales, then I will chase it back again. I booked a hotel in Shrewsbury for last night and will be staying again tonight because I know today will be a packed day of driving. I will need all the energy I can get. The tour actually starts at Bristol Temple Meads but the traction on it at that point, a DB Cargo class 66, isn't as desirable as the traction replacing it at Shrewsbury. The class

289

66 will come off here and two bright-yellow class 97s will join on to the front together. They are practically class 37s but fitted with the modern signalling system ERTMS (European Rail Traffic Management System), for use predominantly on the Cambrian Line, running from Shrewsbury into north-west Wales, as this whole section of track uses the signalling system. ERTMS relies on the driver monitoring a screen within the cab instead of external signals indicating how the train should proceed. It allows trains to run closer together due to constant feedback of relative distance between trains, instead of indication of proximity at every external signal. Network Rail installed four class 37s (37100, 37170, 37178, 37217) with ERTMS, redesignating them TOPS numbers 97301, 97302, 97303 and 97304 respectively. The designation

Class 66

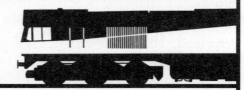

Build date 1998–2008

Total produced 480

Number in service/preserved 327

Prime mover EMD 710 two-stroke diesel, Class 66/9: 12N-710G3B-T2, Remainder: 12N-710G3B-EC

Power output 3,300bhp

Maximum speed 65mph and 75mph

Current operators Colas Rail, DB Cargo UK, Direct Rail Services, Freightliner, GB Railfreight

Nicknames Shed

'class 97' has always been used for departmental locomotives, or locomotives used for testing. In this case the class 37s, or rather 97s, were used to test the ERTMS signalling on the Cambrian Coast Line – a better job than some of the previous class 97s have had to do. 46009, a locomotive that would've hauled passenger services previously, was renumbered 97401, although it was never renumbered physically due to its planned short lifespan. It was used to test how robust nuclear flask wagons were by letting it loose on an open stretch of line at 100mph and crashing it into the stationary flask, lying on its side across the tracks, writing the loco off instantly in an inferno of tangled metal, dirt and combusted diesel. 97301–97304 received very different treatment and they now happily wear their pretty yellow paint jobs with high-intensity LED headlights poking through and red buffer casings, contrasting brilliantly with the intense yellow. On the Cambrian Line, now that testing is no longer required, they sometimes run on the timber train from Aberystwyth to Chirk, and they are used for route learning (an important component of driver training) and railtours. They also perform other duties outside the line as most class 37s would.

After watching it depart here at Shrewsbury at 10:17, I will jump on to the A458, roughly following the line, through Shropshire, into Wales, up and down through Snowdonia and connecting up to Afon Mawddach, the river the 97s will cross over at Barmouth. Barmouth Bridge is arguably one of the most beautiful trainspotting locations in the UK. Standing on the northern side, the bridge stretches over a kilometre of water and sand, the ratio of which depends on the tide, with a backdrop of Snowdonian hills. Once across the completely straight bridge, the line veers to the left and inclines, goes through a brief tunnel and continues climbing while elevated on concrete supports above the quaint seaside town of Barmouth – an infrastructural juxtaposition that tickles many different parts of my brain. It stops at Barmouth Station at 12:58 and gives me a chance to get ahead to the next point where I will catch it again, a foot crossing near Llanfair. After that it is on to Llandecwyn Station where I will wait for the class 97s to pass but I will stay in my car. Once they pass, I will safely leave the lay-by and rejoin the road. I have chosen this particular location as the road and the tracks swoop alongside each other passing over Afon Dwyryd, a river not as wide as Afon Mawddach, less jaw-dropping, but just as pretty. After that, I will continue on to Pwllheli to catch the arrival.

The beige and brown mark 1 carriages are already standing at platform 7. The train seems reasonably sparse but approaching the yellow line of the platform I can see all the way down to the end of the train where there is a considerable number of train enthusiasts milling around – the class 97s must've already attached. I walk down to the end. At least fifty train enthusiasts are arced around the locomotives, starting at the edge of 97302 *Ffestiniog & Welsh Highland Railways* and sweeping 90 degrees to the beginning of 97304 *John Tiley*. Both of the 97s are snoring, soon to be awoken by a green signal.

The stewards on the platform shout down to the enthusiasts to get on board as the train is nearly ready to leave. One by one, they peel off from the edge of the arc and scramble down to their carriage. Those who weren't satisfied with previous photos will have a good opportunity now the crowd is thinning out. I stand aside from the crowd, wondering how many of them are just here to watch the departure and how many are riding on the railtour. A moment later I'm amazed to see the crowd around the locomotive has completely dispersed. There are a few others towards the end of the platform but all of those who were huddled around are actually riding on the tour. I touch 97302 briefly and then 97304, their shells rattling, still asleep, dormant, but ready to throw absolute terror into the air.

Three men in hi-vis and another man in a West Coast Railways fleece walk up to the locomotives. Two of them nod as they walk by and the man in the fleece says hello.

'I'm planning on chasing you all the way to Pwllheli today,' I say with a smile.

'I'll make sure to give you a couple of toots if I see you,' the man in the fleece says.

'I'll catch you at Barmouth first though!'

'OK, mate, I'll see you there.'

He must be the driver then. He looks like a nice man, warm energy, seems down to earth. Two of the men in hi-vis get in 97302 and the other one climbs with the man in the fleece into 97304.

I walk down to where a few middle-aged enthusiasts have set up along the green low-level fence that borders the end of the platform on three sides. I'm very surprised not to see any of the younger generation, particularly as the double 97s are quite a rare occurrence. I settle in between two enthusiasts and rest my wrists in the supporting beam of the fence, in between the bars. A worker for Transport for Wales walks down the platform to watch the departure.

'Hello, mate, love your videos,' he says.

I smile. 'Thank you very much.'

Our conversation is cut short as the signal overhead changes its aspect. Everyone shuffles slightly in preparation for departure. Usually the change of the signal would be inaudible, but Shrewsbury is controlled by semaphore signalling that requires mechanical input from the signal operator in the behemoth Severn Bridge Junction signal box that sits in between the Shrewsbury–Wolverhampton Line and the Cambrian Line. As one of the 180 levers is pulled inside the signal box, a series of wires and pulleys translates the mechanical input into a mechanical output over 100 metres away, raising or lowering a side-facing rectangular plate. Horizontal and highlighting red: stop. Lowered at 45 degrees to the horizontal and green: proceed. The dropping of the signal bounces slightly and makes an audible *tick tick*. The class 97s are ready to depart. Brake pressure is released and all hellfire breaks loose.

Location: Snowdonia, 22 May 2022, 12:14

Two V-22 Ospreys blast overhead as I'm driving Lucy through Snowdonia National Park. 'Oh my God, this must be the Mach loop,' I say to myself – a route where air forces train through the tight and winding valleys, often following the road as a guide. Lucy has to drop a gear to maintain speed up the hill. This is completely new territory for both of us. Neither I nor she has experienced elevation as tumultuous as this and landscapes as staggering as the mountains towering around us. They are on the verge of what I'd class as a hill, but the sheer rock faces of some of the high points, where trees cannot successfully grow, look very mountainous to me. Winding up to the top of the crest of a hill, I look in my rear-view mirror to see the route I have completed over the past ten minutes. Looking back at the road ahead, it winds back down the other side. There aren't any more hills, just a gentle but twisty descent to the coastline.

Barmouth is very sunny. Most of the journey has been spent dipping in and out of drizzle. The sea line sucks away the grumpy clouds and leaves a cool blue, making it too hot for my Missoni jumper. I have parked just after the concrete supports that allow the tracks to pass at second-floor level over the sleepy seaside town. The street on which I have parked runs right next to the water with benches and flower boxes placed alternately. A very tight parallel parking space was offered as a challenge. I accepted with an audience of three people sitting outside a café and an elderly couple sitting on a bench. Lucy reversed in at a questionably tight angle but she was just far enough away from the car in front to slide in perfectly with the wheel fully turned to the right. Once settled, I check live.rail-record. The 97s are running fifteen minutes late so I lock the car and stroll off to find some food. A Co-op will do, a simple meal deal, Lucozade orange – God, I'm thirsty too. I look up and down the high street as I approach the T-junction. The Harbour Fish Bar stares at me. Oh sod it, I'm by the sea. It has to be fish and chips! The battered fillets are lined up on the display grill ready to go. I politely request a cod and chips and it's enclosed in a takeaway box with salt but no vinegar (I find vinegar can sometimes soggify the crispiness of it) and two cans of Old Jamaica Ginger Beer too. I bid the folks in the chip shop a good day and head north towards Barmouth Bridge.

Above the brief tunnel between the railway overpass and the bridge, there's a small viewing area. Paved guides lead towards benches with the edges protected by low-level fencing. The side that faces the bridge is packed with train enthusiasts. Some of these guys are different from any I have met before; they are almost like bird watchers in Richmond Park but without the camouflage. The lenses on their cameras are huge! I certainly wouldn't want to disturb them, it looks very serious. Although it would be ideal to get two epic shots: one over the bridge, then run over to the other side and get another shot of the two yellow 97s thrashing above the normality below. I wonder if that's what the photographers plan to do.

Anyway, this isn't the spot I planned in the first place, it just struck me as I was walking by with my warm box of greasy fish and chips. The spot I am heading to is actually a footpath that runs directly alongside the tracks and over the bridge. As the track leads left into the tunnel, the footpath elevates higher than the track but stays the same distance from it, rising to a bird's-eye view. I'm going to find the sweet spot where I'm at the same height as the exhaust pipes of the 97s so I can experience the thrash in the most intense way possible. Scouring YouTube videos last night, I was trying to find the point at which the driver begins to thrash due to the incline into Barmouth. Three videos ranging over the past ten years showed the drivers thrashing once in the tunnel, another showed thrashing just beforehand, so I'm hoping the driver will do the same today. I don't want to touch my fish and chips yet. I won't be able to finish it all now and if I were to have a bit of it, opening the box would release the heat and make them go cold. I will save them for when I'm back in the car.

I can hear noises from the photographers above the tunnel. All of them are looking through their lenses. The 97s are here, and coming at some speed too. Their side profile exposes two bumps of yellow followed by a shiny streak of light, the sun reflecting off the white roofs of the carriages on the other side of the valley. There's a gentle descent towards the bridge, leading around to the left, turning the side profile into an almost head-on position. Speed is almost impossible to gauge at over a kilometre away and with only the increasing size of the front profile to indicate the train's advance. At one point I'm almost convinced that it has come to a stop. I move further up the footpath to catch a different angle. Still moving. Headlights shining bright in the blaring sun. The lashing wind has become quite intense, throwing the tufts of green along the hills into the shade of thick but dispersed clouds passing over them. Muted copper-sulphate-blue sea water is brought to life in the light, flowing in between the repetitious supports of the bridge. All the while the locomotives are cruising closer. Seabirds, unlike any I have seen before, are squeaking frantically in the water below. One of them has caught a fish and the other birds are chasing after it in the hope of nicking it for their own. I stare at them, mildly annoyed that they're interrupting the soundscape — a job reserved for English Electric traction motors at this moment.

97302 and 97304 pass over the two arched overhead supports of the bridge. The arches carry the tension and distribute it to the reinforced supports at either side, allowing small boats to pass under the bridge with a bit more breathing room. 'Here they come!' I say to myself. 304 has just reached the initiation of the left-hand corner towards me. 'Please thrash,' I think to myself. He seems to be carrying a fairly low speed so I imagine he will need to power up soon. I raise my hand to wave, hoping that the driver will remember me from the station.

The first kick of thrash emanates through the wind as

302 follows around the tight-radius corner. It builds up to a constant growl; the turbo whistle is there but the piercing sound of it has been dampened by the wind. Tingles start to flow along my arms. *Du Du Du DIII Du Du Duuu.* The driver lets off a quick succession of tones as the train curves around towards me. 'Thank you!' 304 and 302 are still thrashing. I see him wave in the cab and the other man in hi-vis is smiling. The roofline of the locomotives is just a metre and a half below me. Here comes the exhaust. Rectangular slots in the curved roofline unleash booms of sonic energy from each combustion cycle. *WoooooAAAAAoooow.* I can't help but scream in reaction to the almighty torrent of thrash passing right next to me, vibrating the ground and the railings I'm holding on to. My body shudders as I briefly look into the exhaust. Deep black, caked in soot. It feels like I'm looking into the centre of the earth for a split second as 304 passes by. 302 follows closely behind and I prepare for the same treatment. The thrash isn't as intense, less bassy and a little bit more rattly, but I can't help but squeal anyway. The coaches follow as the

locomotives power down leading into the tunnel. The driver now has sufficient momentum to coast up some of the incline but I imagine he'll have to reapply some power towards the top. It's the first time I have noticed a difference between the sounds of the thrash of 37s. On the railtour, the differences were almost indistinguishable between 37425 and 37218 in terms of the sound they emitted; I thought the bashers were hearing discrepancies at a frequency I wasn't tuned into. Here, listening to 97304 (37217) and 97302 (37170), there's a clear difference. The last carriage passes through into the single-track tunnel with the red tail light flashing as it trails off into the darkness.

I hang around way too long in Barmouth. Sobering up from the exhilaration of the pass over Barmouth Bridge on my walk back to the car, I remember my plan was to immediately run back to the car so I could get ahead of the train as it paused at Barmouth Station. It must've departed by now. With my two cans of ginger beer and my unopened fish and chips, I run down the high street back to Lucy.

'Did you see the yellow locomotives just now?' I ask the same people that were sitting outside the café earlier.

'Yes! They were so loud!' one of the ladies calls back.

'They are my favourite,' I tell them as I hop into Lucy. I do a quick U-turn and head back through town.

Location: Harlech, 22 May 2022, 13:46

Llanfair foot crossing has had to be missed in order to catch up. Passing through Harlech, I'm keeping my peripheral vision attuned to any yellow interferences in the usual green of the landscape. The A496 follows the tracks roughly; I haven't looked but I presume the train is still ahead. Lucy drops down a gear to climb the slight increase in elevation, then changes back up as it plateaus, then I release the accelerator to let gravity take us downhill. Dead ahead, in the break of some trees, I see the two yellow locomotives and the trailing carriages. 'Oh, for goodness' sake.' It's definitely passed Llandecwyn Station now. That must be Afon Mawddach down there.

In a matter of minutes I'm approaching Llandecwyn Station. It looks tiny, its only notable features a shelter from the rain and a lay-by just big enough for Lucy. The road swings to the right and immediately lines up next to the track with a grey flat stone wall dividing the two. I imagine there's concrete reinforcement on the other side to prevent stones and debris from flying on to the track in the event of a car accident. The wall that hides the track is joined by a metal guard rail, like the ones found at the edge of motorways to maximize the safe division between the two directions of travel. To the left, the topography is low and minimal – we are at the mouth of Afon Mawddach. To the right, a sizable hill drops away. On this side there is no wall and I can see the river clearly and the wide sandy area that surrounds it. A little further and the wall to the left drops away. 'Ahhh, this would've been perfect,' I think to myself. Everything is lined up and follows the same gentle left-hand curve. The edge of the pavement on the right; the centreline of the road; the kerb on the left-hand side; the guard fence on the left; the low, non-obscuring wire fence that prevents people from walking on the tracks; the two gorgeous rusty rails and the fence on the other side of the tracks. All lines congruent with one another. Both the track and the road pass over a long, swooping bridge that changes the perimeter fences to slatted metal guides, then back on to land with the previous combination of fences and guard rails. This continues for far longer than I'd expected. In my planning, I thought it was just passing over the water where the road and the tracks run next to each other but it continues in the direction of Porthmadog for half a kilometre, later diverging and going separate ways, marked by a farmer's crossing that has a convenient lay-by just as the unification of the road and rail starts in the other direction. 'I need to come here on the way back and I one hundred per cent need to make sure I'm here first,' I think to myself.

The buzz of excitement for the adventure ahead spurs me on to drive the next twenty-five minutes to Pwllheli.

Location: Pwllheli Station, 22 May 2022, 14:24

I park Lucy in some bays next to the marina. The weather has taken a turn and the wind has started to carry lashing drops of rain. Pwllheli Station has a very confusing entrance. Initially I thought a café was the way in but I carry on around the edge of the building to an open canopy leading on to the platform. The other side of the café opens out into this area with metal tables and chairs. It seems to be quite popular. A single-bay platform highlights the end of the Cambrian Coast Line, some of the most dramatic scenes a track can pass through brought to a halt at a very mediocre station. The utilitarian canopy does not extend further than the concourse, which carries little to no indication that it's a station at all, apart from the double arrow BR logo and Pwllheli

Station sign. The platform is merely there to serve its purpose, that is all. Further shelter is somewhat offered by the rear of Home Bargains but that only lasts for 10 metres, replaced by a cross-linked fence preventing passengers from falling down into the Home Bargains car park.

97304 and 97302 are due in thirty-five minutes, which finally gives me time to eat my fish. Earlier, I gave up hope of keeping the food warm and picked at the chips while driving. They were surprisingly nice, having held their crispiness and a perfect level of saltiness through the eventful moments since leaving Harbour Fish Bar. The crispy cod was difficult to eat in the car so I thought I'd save it till Pwllheli. I plonk myself down on the rough tarmacked floor and get stuck in, shamelessly picking the fillet up with my hands and taking bites out of it. It's cold now but will have to do as it's the only food I've had today.

Both 97s trundle into platform 1. I have taken refuge under the canopy, the rain now relentlessly washing up and down the platform. Once at a complete stop, the bashers are unloaded on to the platform. We have two hours in Pwllheli before the return journey commences. Photos are taken and the two 97s reverse back into the sidings to allow the locomotives to detach and run around the carriages to reattach on the other end and reverse back into the station. I head back to the car to have a moment to relax before the three-and-a-half-hour journey back to Shrewsbury.

Departing the station won't be particularly exciting to watch as the locomotives will be so far forward and there isn't a good vantage point to see them from after the platform. A level crossing just down the road seems to be a safe bet so I make my way down there. Lucy stays in the space I left her in. It's a regrettable decision as, halfway along the marina wall, I realize the walk is a lot further than I expected and I am too far away now to head back, pick her up and drive to the level crossing before the 97s leave, which is in five minutes. I will need to leg it back in order to beat it to the farmer's crossing after Porthmadog to run alongside it.

Two families with young children are standing on the pavement next to the level crossing, presumably waiting to see the departure of the vintage locomotives. They may interfere with my video so I will have to think of a secondary option; also, I have just realized that the siren of the level crossing will be ding-donging away before the approach of the locomotives so that too will take away from the moment. Both the road and the track cross over a river leading into the marina. The track stands on a bridge slightly higher than the road and there's a metre gap between the bridges where, before the water, a strip of grass runs along. To create a boundary here, a waist-high wall connects the road and the railbridge together, offering a perfect but slightly precarious viewpoint. I'd have to hold on to the side of the buttress to mitigate any chance of me falling down into the water. I climb up as a test. It seems good. It'll do.

A two-tone from 97302, which is now leading 97304 followed by the rake of mark 1 carriages, cries out from Pwllheli Station. There is no thrashing, just a gentle pull away. I stand up on the wall. The level crossing starts to chime, the wind making it barely audible. Cars stop for the closing gates and traffic builds up. With drivers' attention no longer on driving, they observe their surroundings. A van stops next to me in the queue of cars. The window rolls down. 'Where's your anorak?' calls out the driver, then rolls up his window. A typical comment on the stereotype of trainspotters, but he isn't wrong. Where is my anorak indeed? It is still spitting with rain, and the man might actually have been concerned about the water resistance of the jacket I am wearing. I can't quite decide what his intentions were but I side with my immediate impression.

97302 wallows along the infrequently used track. Still no thrashing. It approaches the bridge. *DU DU DU DUDU* – a quick succession of tones rattles out from the horns. They sound a lot more raspy and powerful than 97304's tones earlier. That's one of the fantastic things about class 37s: their horns are so characterful and almost unique to each locomotive. Not only that, but the horns are different on each end of the locomotive too, and it's possible in some 37s to do tones from the rear while in the opposite cab. This is a feature that was used for indication of movement in yards so the driver didn't have to change ends. 302 rumbles by next to me, followed by 304. As 304 is midway past me, the rear horn of 302 toots! They must be in the rear cab of 97302, or perhaps the driver in the West Coast Railways jacket did tones from the rear for me. Brown and beige coaches whoosh by, some of the passengers pointing at me, my eyeline at the lower edge of the window. I jump down and start running back to the car. I cannot let it beat me again.

No sign of it on the whole journey. We've just gone through the outskirts of Porthmadog on the A487, a nice long stretch of road. Next, Minffordd, and after that the crossing. I must've passed it on the way, surely. I beat it to Pwllheli when it had a head start at Barmouth on the outbound journey so surely the head start I gave it at Pwllheli for the inbound journey, being less than Barmouth, should've put me in front. I don't know. Too many variables. I'm starting to get anxious at the thought of missing the 97s again at that awesome spot.

Yellow in my rear-view mirror. Oh my God! I must've passed them just now while they were concealed by the bushes. This is close. It needs to slow down considerably through Minffordd and it's quite windy. For me and Lucy, Minffordd will rein us in too. This will be incredibly tight. I need to get to the crossing first.

Location: Minffordd, Wales, 22 May 2022, 17:01

We're here. I pull Lucy into the gravelly pre-crossing zone for the tractors, avoiding parking where the tractors would need to stand while the gates opened, just on a grassy patch. Engine off. Immediately, I pull out my phone and check live. rail-record. It hasn't passed through Minffordd yet. 'OK, we can relax, Lucy.' I tap the wheel in reassurance.

A gust of wind hits me the second I open the door. It's such a beautiful landscape. Drizzle softens the colours of the deep green trees on the Minffordd side of Afon Mawddach and continues across to the other side, making the bold hills turn hazy. The low-level cloud rushes by in the wind, low-hanging wisps brushing the top of some hills. Other, taller hills are engulfed altogether near the top. Sheep graze in the field over the crossing, enjoying the rich green grass that disappears in patches, interrupted by dark brown shrubs that sit in puffs 20 centimetres above the grass. Looking to my left, I follow along the track to Llandecwyn Station where the road departs from the unification. It's a seriously far distance and along this line the locomotives should only be going around

30mph, so when the speed limit on the road changes to 40mph I should be able to catch up. Nevertheless, I will have to get in my car and go before the 97s pass the crossing in order to accelerate to the 30mph speed limit and keep level with the 97s. I am feeling very nervous but I am ready.

I have spent the whole day in my own company. I have just realized it. Something that before, when I was a teenager, would have brought on feelings of loneliness. Wandering around the corridors at school, I felt alone and slightly depressed. Heading home I would still be physically alone but I would find company and comfort in the world I'd created on my 2-foot-by-8-foot board, with its miniature tracks and proportionally miniature trains. Right now I am alone, and have been for over thirty-six hours, but in the same way that I used to feel with my model railway: I am totally at ease, calmed by the situation and happy in my circumstances. I am alone, but I am certainly not lonely.

97302 followed by 97304 rolls down the descent, approaching sea level. Light brake application is holding the consist from rolling at too fast a pace. 302 has a fantastic yellow snow plough to match the paint job, unlike 304. The composition is beautiful, the grazing sheep matching perfectly with the cream of the carriages, the swaying trees, the speeding clouds and the bright yellow locomotives. My breathing gets heavier and heavier; I start to laugh nervously.

They pass a small halt 75 metres away from the crossing where a few spotters are standing. *DI DUUUUWUWU DU DU*. Tones! I laugh and stick my thumb into the air in appreciation. The throatiness of the tones reminds me of 37510's low tone at Gloucester Station early on that winter morning. 97302 rises like an undersea volcano breaking through the crust and spewing molten rock. Brakes are released and the engine is notched right to the maximum. Its turbocharger scream spools up and makes me want to scream back in an equal and opposite reaction. The English Electric type 3 engine is chaotically convulsing the environment around it, pumping out energy at every turn of the crankshaft, energy that I engross myself in and release in my own way. Laughing and buzzing, I could jump, I could scream. I need to get into my car and go, but I can't. I have to hear more thrash. 302 passes with a wash of intense thrash followed by 304, more bassy and more aggressive. Hysterical laughter takes over my body as I realize I might've messed up, having waited too long for it to pass. I don't care, I'm wrapped in thrash and the serotonin is not leaving.

The keys are in the ignition ready for me to go. 'Quick!' I throw open the door and get in. Seat belt on. 'Bugger!' Turn the key. 'Come on!' Foot on the brake. Into drive. Handbrake off. Look over my shoulder. Look forward. Look over my shoulder. Look forward. All clear. Foot off the brake. Accelerate gently off the gravel and on to the road. Accelerate up to 30mph. Don't get distracted by the train. The carriages that were flowing by are now decreasing in relative speed. We are now level. Bashers on the train point at me from the carriage. I lower my window. A 40mph sign passes. I accelerate up to 40mph, claiming back the carriages that had passed me. The right-hand curve obscures the 97s until we get on to the bridge. There! I wind the window all the way down. 304 is thrashing hard. 'Yeahhhh!' My endorphins max out, bashing the rev limiter of the amount of serotonin possible in my bloodstream. 'YEAHHHH!' There are drivers in the rear of 302 after all, and they've spotted me. Spamming the rear horn of the lead class 97s. *DU DIIII DI DU DI.* 'YEAHHHH!' Almost in reply to each scream, the drivers give me a series

of tones in return. My eyes are on the road. 'YEAHHHH!' My voice becomes hoarse. I think this is the loudest I have ever shouted in my life. I feel like a caveman returning to the building blocks of life. Roaring in celebration. I feel unstoppable. Absolutely enthralled and immersed in the moment. Nothing matters right now apart from the joy and happiness that is swirling around in my mind. A smile stretches wide from one side of my head to the other. 'WOOOOOOOO!'

The powering locomotives start to increase in speed. I take one brief look to my right and see the driver looking at me out of his window with a smile across his face and a thumbs up. We immediately turn back to the road and track in front of us, in that moment one of us driving a sixty-year-old class 97, the other driving a 93hp Nissan Micra, both flying over Afon Mawddach. I hope I was able to transfer some of my happy energy to him, in the same way the locomotives transferred it to me due to his enthusiastic driving.

The road peels away, past Llandecwyn Station. Two tones from the lead powercar to say goodbye. I look in my rear-view mirror and see cream and brown carriages speeding through. I raise my window until it's shut. 'Oh, that was mega!' I can hardly catch my breath. On each exhale I laugh then draw air back in through the gaps between the edges of my smile and my teeth. I maintain my focus on the road. I can hardly believe what's just happened. My hands are shaking.

Pulling over after a few kilometres, I take a minute to calm down and process the moment. Engine off, I sit there going through the series of events on my mind's cinema screen. Thrash. Tones. Joy. I can't stop laughing. The class 97s continue along the coastline. I have plans to catch the train further down the line at Tywyn. I may get it there, I may not. For now I am happy just to sit in my car. I'm not going to think about anything for a moment, just soak it all in. 'Thank you, Lucy.' I tilt my head back into the headrest and hold on to the steering wheel with two hands, then look right towards the sea, the sun reflecting softly off the water, a field of cows in the foreground. I think I can just make out the train in the distance. I think that's it. A black calf feeds from its mother near the fence, another one, alone, stands there looking at me. I wind down the window and smile and say hello. It raises its head and nods, shakes off some flies, and walks towards its mother. Shrewsbury is two hours away, or three if I want to catch the 97s again. Either way, I am blissfully content.

Way out →

Trainspotter's Bucket List

❏ Spot an HST along Dawlish sea wall

❏ Spot the *Flying Scotsman*

❏ Spot a class 37 on the mainline

❏ Hear a class 37 thrash

❏ Ride on a class of train on its last day of
passenger service

❏ Ride on a railtour

❏ Ride on a heritage railway behind some classic
diesel traction

❏ Cab ride in a heritage diesel locomotive on a heritage line

❏ Ride on the footplate of a steam engine on a heritage line

❏ Get Ilkley Baht 'at tones

❑ Get 'Imperial March' tones

❑ Get mega tones

❑ See an HST crossover

❑ Take a journey along the English Riviera

❑ Take a journey through the Pennines

❑ Take a journey along the West Highland Line

❑ Spot the Eurostar

❑ Ride the Eurostar

❑ Spot a battery locomotive train on the London Underground

❑ Go to a train scrapyard to see train remains

❑ HST ride from Plymouth to Edinburgh

❑ HST ride from Cardiff to Penzance

❑ Spot at the Glenfinnan Viaduct

❑ Spot at the Ribblehead Viaduct

❑ Ride on a Class 455 from Waterloo to Dorking

❑ Ride to Windsor and Eton Riverside

❏ Spot the Midland Pullman

❏ Spot the Caledonian Sleeper

❏ Ride on either the Caledonian Sleeper or the
Night Riviera Sleeper

❏ Ride on the Isle of Wight Railway

❏ Spot a test train

❏ Ride on all the metro systems in the UK

❏ Spot in Japan

❏ Spot all over the world

❏ Go to Crewe Heritage Centre

❏ Go to a depot open day

❏ Go to the National Railway Museum and touch the
A4 *Mallard*

❏ Spot an A4 at a heritage railway

❏ Go to the NRM Shildon to see 43102, the best class 43

❏ Donate to a heritage group and help preserve our
favourite locomotives

Glossary

750V DC/third rail The electrical power supply for trains, typically situated on the outside of either of the two centre rails.

AWS Automatic Warning System: a safety system that ensures drivers obey signals.

Ballast The stones that surround and secure the sleepers that are attached to the rails, to prevent them from moving.

Barrier coaches Carriages used to connect non-matching coupler types, allowing locomotives to pull trains comprised of various carriages.

Bash To ride behind a desirable locomotive for the pleasure of experiencing the sounds, racking up the miles and the thrill of being dragged by a beast.

BMAC A rail and road headlight manufacturer.

Bogie The assembly connecting the wheels to the frame of the vehicle, consisting of the wheels, suspension and brake components.

Bowled When the view of a desirable train or locomotive is blocked by another train.

British Pullman A charter train operated by Belmond. Stationed at Stewarts Lane Traction and Rolling Stock Maintenance Depot and running from London Victoria to destinations across the country, it is the most luxurious train in the UK.

Buffers Used to stop trains from going off the end of a line. Also, compressive parts at the ends of railway vehicles that cushion movement transferred between other connected vehicles.

Carriage A railway vehicle that carries passengers.

Chopper Nickname for a class 20.

Clag Fumes that billow from the exhaust of a locomotive, usually very sooty.

Class The number that identifies a particular locomotive series.

Consist A collection of rail vehicles coupled together, otherwise known as a train.

Convoy A train consisting of a locomotive hauling other locomotives.

Crompton Nickname for a class 33.

Deltic Nickname for a class 55.

DMU Diesel Multiple Unit.

Duff Nickname for a class 47.

EMD Electromotive Diesel.

EMU Electric Multiple Unit.

EWS England Wales Scotland.

FD Freight Depot.

Firebox Where coal is burned to boil water that creates steam to power a steam locomotive.

Footplate Where the driver, second man and fireman stand on a steam locomotive.

Gen Information discoverable in railway enthusiast communities, regarding movements or events on the railway.

Grid Nickname for a class 56.

Gronk Nickname for a class 08 or class 09.

GSM-R Global System for Mobile communications-Railway.

Headcode A code used to identify a train.

HST High Speed Train, typically associated with the class 43.

Junction A divergence of a route or line.

L.I.P./LIP Locomotive Inspection Point.

Line speed The maximum speed at which a train can travel on the line, governed by safety, not the terminal velocity of the train.

Locomotive A powered railway vehicle used for pulling trains.

LSL Locomotive Services Limited.

Midland Pullman A charter train operated by LSL: a bright blue HST that runs around the country, based on the original Midland Pullman from the 1960s.

MPV Multipurpose Vehicle.

NMT New Measurement Train.

Paxman Valenta The first engine fitted to the class 43 HST.

PLPR Plain Line Pattern Recognition.

Powercar A locomotive used on either end of coaching stock, typically with one cab.

Railhead The part of the rail that makes contact with the wheel of the train.

RHTT Rail Head Treatment Train.

RTC Railway Technical Centre.

Shed Nickname for a class 66.

Shoebox Nickname for a class 73.

Shunter A locomotive used to move trains around marshalling yards or depots. Also, a person whose job it is to oversee these movements.

Siding Track where trains and carriages can be stored or marshalled.

SITT Snow and Ice Treatment Train.

Station Where trains stop in order to let passengers off and on or to transfer goods.

TC Terminal Complex.

Terminus A station that is at the end of a railway line.

Thrashing When a locomotive reaches maximum power to haul the train, usually chucking out clag and making lots of noise.

TMD Traction Maintenance Depot.

TOC Train operating company.

Tones Sounds made by locomotive horns.

TOPS A railway computer system used to process the locations and operational status of locomotives and trains.

TPWS Train Protection Warning System.

Tractor Nickname for a class 37.

Train A series of connected railway carriages or wagons moved by a locomotive or by integral motors.

TRSMD Traction and Rolling Stock Maintenance Depot.

UML Up main line.

UTU Ultrasonic Test Unit.

VP185 One of the engines that superseded the Paxman Valenta in the class 43 HST.

Whistle boards Trackside indicators for when drivers should sound their horn.

Acknowledgements

Thank you to my parents, Jenny and Seamus, for giving me the creative tools that have made this happen. Thank you, also, to my brother, Ben, for being a great companion since the early days.

Thank you to my wonderful girlfriend, Amy, for being such a great supporter and for always being there for me. The class 455s will forever bring happy memories of us together.

Thank you to all of my friends and companions on the adventures I recount in this book. Ryan, Gordon, Benji and so many more – you're a huge part of what makes trainspotting such a joy.

I also want to thank my editor, Henry Vines, for your guidance and vision. I would also like to thank the team at Penguin Random House who have brought this book to life: Danai Denga, Katrina Whone, Barbara Thompson and Dan Balado for your editorial work; James Alexander and Phil Lord for your brilliant designs; Catriona Hillerton for your amazing production work; Richard Ogle for your beautiful cover art direction; Becky Short and Tom Hill for your PR expertise; Emma Burton and Rosie Ainsworth for your incredible marketing; and finally Oli Grant, Roy McMillan and Christopher Thompson for all of your help with the audiobook.

I'd like to thank Isaac Marley Morgan for your beautiful photographs for the book cover.

Thank you to my agent, Millie Lean, for reaching out to me, having the vision of trainspotting becoming a book and for the support throughout the journey. Thank you to my team at YMU: Amanda Harris, Elise Middleton, Naomi Tunis, Anna Dixon and Aram Fox.

Thank you to everyone who has ever supported me and watched one of my videos. I am truly overwhelmed by your kindness and positivity.

Picture Acknowledgements

All photographs are the author's own. Original illustrations have been created by the author and James Alexander, with thanks to the following sources and references:

Creative Commons licence images: p. 45, Andrew Rabbott; pp. 51 and 249, Geoff Sheppard; p. 74, train_photos; pp. 212 and 280, The Basingstoker; p.228, Bs0u10e01; pp. 254 and 262, Ruth Sharville.

Shutterstock images: p. 73, image no. 1632755761, Cristian Paris; p. 157, image no. 195441611, Tom Saga; p. 171, image no. 397871773, Arcansel; p. 268, image no. 1234625242, richardjohnson.

The British Rail double arrow is a registered trademark in the name of the Secretary of State for the Department for Transport.

Francis Bourgeois is an English trainspotter and social media personality, best known for his videos posted on Instagram and TikTok on the topic of trains.

He began making TikTok videos on trainspotting in 2020 and has received widespread acclaim for his joyful and uplifting content.

He has collaborated with public figures including Thierry Henry, Joe Jonas, Rosalia and Sam Fender.